Praise for *The Wood Age*

'Stunning... nothing less than a complete reinterpretation of human history and prehistory... written with enormous verve and pinpoint clarity... No review can match the richness of Ennos's book.'

The Sunday Times

'Passionate... In this very readable historical survey, Ennos argues that not only do we need to reassess the role wood has played in our history, but by embracing a new age of wood, we can help to reset our broken relationship to the natural world... [A] fascinating "wood-centred view" of our history.'

Guardian

'Ennos... a specialist in the mechanical properties of trees, shares his insatiable curiosity with us. He applies his sharp eye for details, and he does so entertainingly.'

Washington Post

'Ennos's special love and concern is for things made from trees... The principles of every significant technology, from tree-felling and carpentry to shipbuilding and papermaking, are described with a precise, almost mesmerising detail.'

New York Times Book Review

'[Ennos] takes a fresh look at the familiar substance, wielding it like a wedge to pry open our past, examine our present and even glimpse our future.'

Wall Street Journal

'A lively history of biology, mechanics and culture that stretches back 60 million years... A specialist in the mechanics of wood, Ennos has a fierce love for his topic.'

Nature

ALSO BY ROLAND ENNOS

The Wood Age: How Wood Shaped the Whole of Human History

Trees: A Complete Guide to Their Biology and Structure

THE SCIENCE OF SPIN

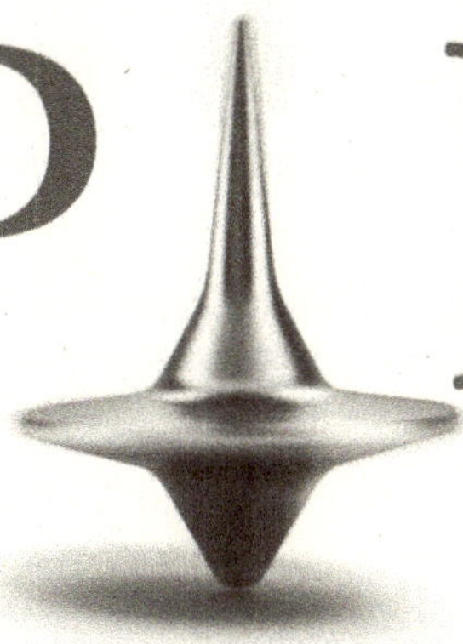

The Force Behind Everything –
From Falling Cats to Jet Engines

ROLAND ENNOS

ONEWORLD

A Oneworld Book

First published in Great Britain, the Republic of Ireland and Australia
by Oneworld Publications, 2023

A CIP record for this title is available from the British Library

ISBN 978-0-86154-671-8
eISBN 978-0-86154-672-5

Printed and bound in Great Britain by Clays Ltd, Elcograf S.p.A.

Oneworld Publications
10 Bloomsbury Street
London WC1B 3SR
England

To Stephen, Catherine, and Amy,
and all of my lovely acquired family

Contents

PROLOGUE

Camels in a Spin

On July 24, 1917, Flight Sub-Lieutenant Sidney Emerson Ellis of the 4 Naval Squadron took off on a routine training flight in Britain's newest and most powerful fighter aircraft, the Sopwith Camel. He banked to the right, dived, and inexplicably plunged, spinning into the ground. It was the first of many fatal crashes, as young, inexperienced pilots took their first flights in the plane, a replacement for its predecessors, the Sopwith Pup and Triplane, which were no match for Germany's new Albatros DIII fighter. With the prospect of losing more pilots and planes through accidents than enemies they could destroy, the authorities plainly had to work out what was going on and seek a remedy.

The cause quickly became apparent: the gyroscopic effect of the Camel's 130-horsepower Clerget rotary engine. In these power units, the whole engine, nine radially arranged cylinders, was attached to the propeller and span around a stationary crankshaft. This apparently bizarre design had several advantages over conventional fixed engines: the cylinders were cooled by the surrounding air as they spun around, so the engine did not need a radiator or heavy water-cooling system; and oil injected at the crankshaft automatically flowed outward, so there was no need for an oil pump. Rotary engines therefore produced more power for their weight than conventional piston engines. But there were side effects. Together, the engine and propeller acted as a heavy gyroscope; they resisted being rotated at right angles to their axis, so that when a pilot applied turning forces on his controls it had unexpected effects. A turn to the left raised the nose of the plane, slowing it down, while a turn to the right forced the nose of the plane down, speeding it up. It was this that caused the uncontrolled descent into a spin that had killed Lieutenant Ellis.

As we shall see, this effect was totally predictable from the laws of physics, but until then it had been ignored, as it had not posed a significant problem to aircraft. Earlier planes, even Sopwith's own Pup and Triplane fighters, had also been powered by rotary engines, but these were lighter units that had spun more slowly, and the aircraft themselves had been designed to be inherently stable. Indeed the Sopwith Pup was described by its pilots as the "most perfect aeroplane ever built." But the stability of early aircraft reduced their manoeuvrability. To obtain better fighting qualities, the Camel had been designed to be more unstable and consequently more manoeuvrable. The top wing was straight, rather than rising towards the wing tips as was more usual, which would have prevented it righting itself automatically, while 90 percent of the aircraft's weight was concentrated in the first seven feet of the fuselage, reducing the torques needed to turn it. The result was a brilliant fighter, but one that could be lethal to inexperienced pilots. By this time, however, there was little the authorities could do to rectify the situation. Pilots were banned from turns to the right at altitudes below one hundred feet, and training machines were modified to hold an instructor sitting behind the pilot who could take over the controls if necessary. With these measures in place, accidents were reduced and the Camel went on to become the most formidable fighter of the war, destroying 1,289 enemy aircraft, and even accounting for the Red Baron, Manfred von Richthofen, himself. The aircraft continued to be a menace to novice pilots, but experienced ones learned to make use of its unusual characteristics, turning to the left in dogfights, for instance, by turning 270 degrees to the right.

The Sopwith Camel. The propeller is attached to the spinning rotary engine, and the craft's weight is concentrated at the front of the fuselage.

This historical crisis is just one example that shows how unaware people are about the science of spin. The mathematics governing rotational motion had been known for 150 years when the Camel first flew, yet even so brilliant an engineer as Herbert Smith, Sopwith's chief designer, had failed to predict how its rotary engine would affect its handling characteristics. And it has largely been fortuitous that gyroscopic effects have rarely posed problems for more recent aircraft. The rotary engine reached the limits of its development in 1918, with Bentley's 150- and 200-horsepower units, after which it was abandoned. After that, propeller-driven aircraft were powered by engines with stationary cylinders—radials, in-line, or Vee—and the gyroscopic forces of the propeller alone have been manageable. Even in jet engines, in which the turbine blades rotate at high speed, the gyroscopic effects are small compared with the aerodynamic forces on the wings, because most of the weight of the engine is concentrated near the centre of rotation.

But there is no reason to feel complacent. Sixty years after the Camel crisis, another affair showed how spin continues to baffle even the greatest minds. As a schoolboy I was myself present at the 1974 Royal Institution Christmas Lectures for young people, given that year by the eminent British engineer Eric Laithwaite, the pioneer of the linear electric motor and high-speed maglev trains. In the fourth lecture, "The Jabberwock," Laithwaite demonstrated to a rapt audience some of the seemingly magical properties of gyroscopes: the way they swiveled around their support without any apparent force being applied; and the way they seemed to hold themselves up, defying gravity. Laithwaite even went on to claim that gyroscopes broke the laws of physics! As you might expect, these claims set off a storm of protest from the physics establishment, and prominent scientists were quick to denounce Laithwaite.

And confusion about spin continues to reign even to this day. Witness, for example, the recent media storm surrounding the observation by a Russian cosmonaut on the International Space Station, Vladimir Dzhanibekov, that in zero gravity a spinning wing nut flips its orientation by 180 degrees every few seconds. This observation caused wonder and consternation in equal measure all across the globe. The press pounced on this so-called Dzhanibekov effect, and the physicists who they invited to comment on the phenomenon were unable to explain why it happens. The Russians were even fearful that such an effect could happen to the spinning earth. If it flipped over in the same way as the wing nut, there would be catastrophic consequences for life on our planet. As we shall see, this is not an isolated occurrence; many other everyday phenomena, from the behaviour of spinning tops to the way children pump playground swings, are subject to abject misinformation on the internet and in physics textbooks alike.

This confusion is particularly unfortunate, because the science of spin pervades all aspects of the world around us. It helped form the universe, shaped our solar system and galaxy, and controls how they behave today. Spin is responsible for shielding the earth's atmosphere from harmful rays, so enabling life to survive on our planet. It shapes our climate and weather, from the periodic return of ice ages, through the global pattern of trade winds, to the local formation of depressions and hurricanes; consequently, it also shapes the ecology and distribution of life on our planet. Most of the machinery that has underpinned the progress of our civilization exploits spin: from spindles, gears, and flywheels that keep it moving; drills and lathes that shape our artifacts; pumps and mill wheels that raise and extract energy from water; and propellers, turbines, centrifugal pumps, impellers, to electric motors that power the modern world. Most important of all, our bodies are systems of rotating joints and levers that are controlled by our unconscious brain. They produce the complex movements of our bodies that enable us to stand up and move about; brandish tools; throw projectiles; and play a whole host of sports.

My aim in this book is to bring clarity to the fascinating subject of spin, so we can see just how it controls the way the world works. Avoiding the mathematics that scientists so often rely on and hide behind, I will provide readers with intuitive physical explanations to explain the mechanics of rotation. Whenever possible I use explanations from the scientific literature, but these are all-too rare; in some cases I have had to devise my own explanations and arguments. I believe this approach should be helpful even for physicists who have long since mastered the mathematics of spin. It should help them dispel whatever doubts Laithwaite brought up about the laws of physics; banish fears that the Dzhanibekov effect could cause a global catastrophe; and help them communicate better with us mere mortals. It should help explain the workings of the world about us. It should shed light on the technology that has built the modern world, technology that was developed long before scientists had anything useful to say about how it works. And this book should help biomechanics and sports scientists to cut through the complexity of the human body to get a better grip on how we move; help design better prostheses and robots; and help sportspeople achieve better performances. And for everybody it should bring the delights of revelation, equipping us with a better understanding of the world about us: one in which spin assumes a more central role. At last this will enable us to appreciate the advantages of our complex jointed bodies and see how they give us a flexibility and economy of movement far superior to wheeled vehicles.

Most of all I hope to return readers to the childlike delights of playing with spinning tops, throwing and catching balls, and swinging sticks around our heads, and show the unlikely links between tightrope walkers and tyrannosaurs, catapults and cricketers, gyroscopes and gymnasts. And if we can understand the mechanics of our bodies, our technology, and the cosmos at large, we will finally be able to understand what really makes the world go round.

PART I

SPIN AND THE WORKINGS OF THE WORLD

CHAPTER 1

How Spin Created the World

> Invention, it must be humbly admitted, does not consist in creating out of void, but out of chaos.
>
> —Mary Shelley

It seems to be a human instinct to want to know more about our own origins: how our parents met; how our country came to be founded; how humans emerged and managed to dominate the world; how the earth was made; and even how the universe itself came into existence. So strong is this instinct that we tend to make up all sorts of stories to explain aspects of the past of which we have no personal experience. In the Bible narrative, for instance, explaining how we came to be here was all so simple. God created the heavens and the earth, with the earth at the centre, and the sun, moon, and stars rotating around it and lighting up the sky. He then moulded our planet into a home fit for his ultimate creation: humankind. He separated the land from the sea, covered the land with plants, and created a host of fish to fill the seas, birds to fill the skies, and animals to live on the dry land. He made it into a perfect place for us humans to live in, and he did it all in the double-quick time of six days.

Today, of course, we know a lot more about the universe in which we live, and about the planet we live on, and consequently we know that we are far from being centre stage. The universe does not revolve around us at all. Instead, the earth is just one of eight planets, several minor planets, and many asteroids and comets, all of which orbit around our sun. And our sun is itself just a minor star, one of hundreds of billions of stars revolving around the centre of our galaxy. And in turn our galaxy is just one of an infinite number of galaxies that make up our universe. But the fact remains that the earth is a great place to live. Light from the sun keeps us warm, and provides the energy that plants use to make our food, while the earth's magnetic field protects us from damaging solar rays. Our seas are rich in salts and nutrients and full of life, and gentle tides caress the shore. The air is easy to breathe and its light winds carry soft refreshing rain to the land, watering our crops and filling our lakes and rivers. We might well agree with Voltaire's Dr. Pangloss that we live in the best possible of all worlds. And as I hope to show in this first part of my book, we owe it all to a motion to which people rarely give more than a few minutes attention: spin.

The first thing that science has had to explain is how our solar system was formed. And if you look at an orrery—a clockwork model of the solar system—you will immediately see clues. The planets all circle the sun in the same plane, and they all orbit it in the same direction. Not only that, but they almost all spin in the same direction, and the moons that orbit the planets rotate about them in the same plane and in the same direction as well. This uniformity demonstrates that the solar system must have been shaped by a single simple process, and all the evidence shows that, like Mary Shelley's Frankenstein, the order was created not out of void, but out of chaos. The generally accepted account of the formation of the solar system is that given by the nebular hypothesis, first proposed in the eighteenth century by the Swedish theologian, philosopher, and mystic Emanuel Swedenborg and the great German philosopher Immanuel Kant.

According to the nebular hypothesis, the earth was formed from a huge cloud of gas and dust. About 4.5 billion years ago, this was hit by the shock wave resulting from a supernova: an explosion produced by the sudden collapse of a large star. This explosion caused the cloud to densify and swirl around in vortices, like the eddies you see on either side of your spoon when you stir your cup of tea.

At this point in our tale, it is worth taking a little time to consider what rotation and spin actually are. After all, they will be central to the rest of this book, and to most of us it is not immediately clear what is going on in these complex motions. It took the genius of the great seventeenth-century scientist Robert Hooke to define them. A particle that is rotating about a central point has two components to its motion. It moves at a constant speed, but its velocity is continually changing because it is also accelerating inward. To keep an object rotating, you therefore have to provide an inward *centripetal* force. And as a consequence of this acceleration, the rotating particle exerts an apparent outward *centrifugal* force that resists it being drawn farther into the centre. An important aspect of steady rotation is that because the force is at right angles to the motion, no energy is needed to keep it going; in the absence of friction a ball rotating at the end of a rope, or a planet orbiting the sun, will keep on moving around forever. In an object that is spinning, exactly the same thing is happening, but since each part of the object rotates at the same rate, the parts that are farther away from the axis of rotation move faster; and as the object spins, centrifugal forces tend to stretch it outward.

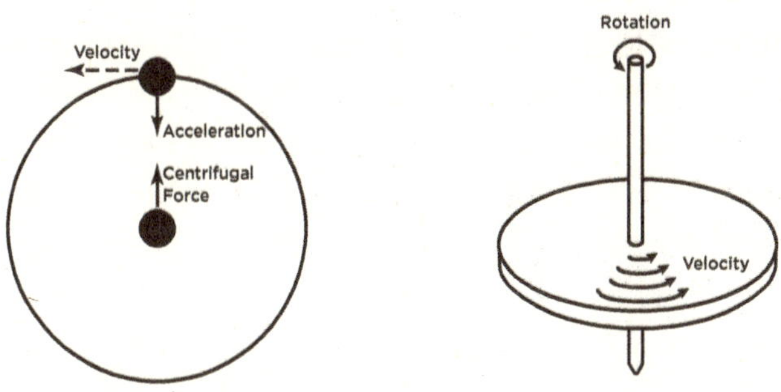

Rotation and spin. In an object rotating around a fixed point (*left*), such as a planet orbiting the sun, its motion is a combination of a constant velocity and an inward, centripetal, acceleration. The earth exerts a corresponding centrifugal force on the sun. In a spinning object such as a top (*right*), the velocity of each point increases with its distance from the centre of rotation.

Just as you need to apply a force to change the velocity of a particle, to speed it up or slow it down, you also need to apply a turning force, what is known to scientists as a torque or moment, to a rotating object to make it spin faster or slower. And just as you need to apply a greater force to accelerate a more massive particle, you need a greater torque to change the spin rate of a bigger, more massive rotating body. In fact, the parts farther away from the centre of a rotating body need more torque to accelerate them, both because they move faster for a given spin rate of the body, and because they are farther away from its centre. The rotational equivalent of mass is known as the "moment of inertia" of a body, which not only takes into account the mass but also how far material is from the axis.

Knowing this helps to explain what should happen to a spinning ball of gas. You might expect that the force of gravity would inexorably draw it into a single flaming ball of material—a new star. Certainly, gravity can easily draw all the particles towards one another parallel to the axis of rotation, flattening the cloud. However, it would not be able to move them all the way inward to the axis of rotation because the centrifugal force of the particles would resist gravity; the gravity could only provide the inward force needed to keep the particles travelling in a circle. Consequently, gravity would merely compress the cloud into a flat disk of rotating particles. The particles would continue rotating about the centre of the disk, like the rings of Saturn. Once there, however, gravity between the particles would gradually draw them together into larger particles, draw the larger particles together into rocks, and draw the rocks together into bigger and bigger boulders. The gravitational energy would heat them up as they collided, and finally, as the giant boulders collided together, this would melt the rock, allowing them to coalesce into spherical planets. The result would be what we have in our present solar system: a succession of planets, all of which orbit the centre of the system, in the same plane, and all of which orbit in the same direction. The apparent chaos of swirling gas would have been transformed into the order of the orbiting planets.

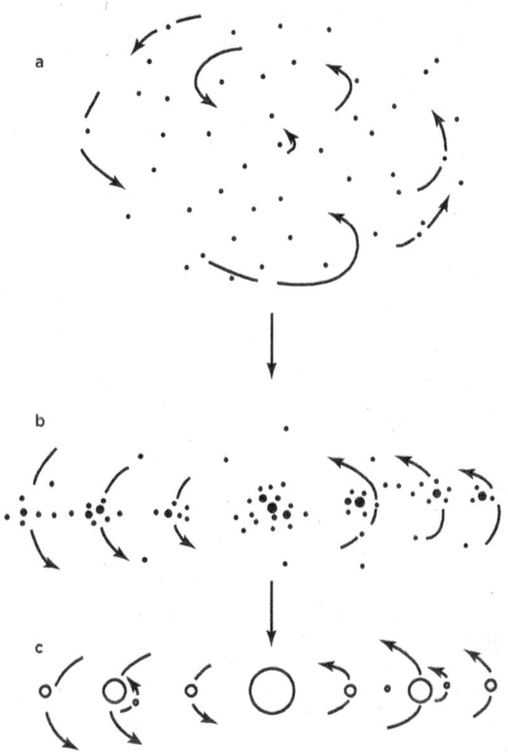

The nebular hypothesis of the formation of the solar system. In its modern guise the premise is that the system was formed from a cloud of spinning gas (a). Gradually, gravity flattened the cloud into a disk (b) and the particles began to coalesce into larger and larger boulders, and eventually into the sun, planets, and moons (c).

The nebular hypothesis not only explains why the planets all orbit in the same plane and in the same direction. It also explains why the planets spin in the same plane and in the same direction as they orbit, and why the smaller objects that orbit around the planets as moons circle them in the same direction. If a small particle was drawn inward by the gravity of a growing planet, so that it approached the planet from a slightly more distal orbit, it would speed up as it moved nearer the centre of the solar system and would eventually hit the planet a glancing blow, causing the outside of the planet to accelerate forward. In contrast, a particle drawn outward towards the planet from an inner orbit would slow down and hit it a glancing backwards blow to decelerate the inside of the planet backwards. Both types of collisions would spin the planet in the same direction: forward. In the same way a moon that was captured by a planet would always rotate forward around the planet, whether captured from outside or inside its orbit. And, in the near vacuum of space, once a planet was set spinning and a moon was set orbiting, they would continue spinning and orbiting indefinitely.

The nebular hypothesis therefore provides a convincing explanation for why the planets move in the ways they do, and it also explains other aspects of the structure of our solar system. In the hotter inner areas of the solar system, only dust would be able to coalesce, explaining why the inner planets are all rocky. In contrast, in the cold outer recesses of the solar system, gases can also condense and freeze, explaining how gas giants such as Jupiter and Saturn were created. However, in some ways our solar system does not resemble the one that computer simulations predict should have been created. Mars seems too small, and the gas giants are farther out than where they would have formed. The reason is probably that the planets are not only attracted by the gravity of the sun, but one another's gravity, too; as a consequence they can alter one another's orbits so that over vast periods of time their behaviour can be chaotic. The grand tack hypothesis suggests that billions of years ago Jupiter may have moved first inward and then outward, acting like a giant wrecking ball, forming the asteroid belt in the process. And collisions between other celestial bodies might also have altered their paths. Neptune seems to have been knocked out of its original orbit so that it now spins on its side.

The earth, too, is unusual. Other planets have relatively tiny moons, which formed around them in just the same way that the planets themselves formed: from clouds of gas and dust. In contrast, our own moon, at one-sixth of the earth's diameter, is exceedingly large. Modern thinking is that the early earth, Gaia, probably acquired its moon following a collision with another planet the size of Mars, Theia. The two planets fused to form a single larger planet, while part of the material broke away and coalesced to form our large moon. This collision would have speeded up the spin of the earth so that it rotated once every six hours or so, but it also gave the earth its unusual tilt of around 23.5 degrees from the plane of its orbit, while the moon settled into an orbit that is tilted around 5 degrees from the earth's orbit. As we shall see, these details of how the earth spins and how our oversize moon orbits around it have proved crucial in making our planet an ideal place to live.

Despite the complexities, therefore, it is clear that spin was, along with gravity, the major factor in forming the system of planets that circle our solar system. However, the most difficult aspect of the creation of our solar system to explain is how the sun, the source of all our energy, and the part of the solar system that contains almost 99.9 percent of its mass, formed at its centre. Since the central part of the cloud of gas would be spinning just like the material farther out, you might expect that gravity would only have been able to flatten it into a series of ever larger planets towards the middle of the system—planets that would rotate rapidly around the centre. Gravity would simply not have been able to have drawn all that matter into a body that is only 865 thousand miles (1.4 million kilometres) across, a hundred times smaller than the orbit of the nearest planet, Mercury. Something must have happened to slow the rotation of the gases at the centre of our solar system, to allow the atoms to spiral inward to form the sun.

Once again, it is believed that the mechanism by which this was achieved involved spin. As the material near the centre of the solar system was flattened, the gravitational energy would have been converted into heat, which would be great enough to turn the gas into a plasma of charged particles so hot that it would start undergoing nuclear fusion. These changes would have two consequences. First, the nuclear fusion would cause large numbers of the charged particles—electrons, protons, and helium nuclei—to be spat out of the core. On its own this would not slow the spin down. However, another consequence of being composed of a plasma would be that the spinning charged particles would set up a huge magnetic field around the sun, a field that itself spun around the core. Trapped within this rotating field as they were ejected, the particles would not move straight outward from the centre of the sun, but be pulled into a spiral motion, travelling along the rotating magnetic fields before finally being flung forward as they escaped, like water being released from a rotating crop sprayer. This process would slow the remaining material down, taking away some of the sun's angular momentum, and allowing gravity to draw the remaining gases farther inward. In turn this would further increase the sun's core temperature, ensuring that fusion reactions could continue to produce energy. It is this energy, released in the form of the electromagnetic radiation, mostly light, that heats up our planet to its balmy temperature and powers the photochemical reactions that algae and plants use to make the food on which we rely.

We therefore have a convincing theory about how our solar system was formed and how it works today. And in recent years astronomers and physicists have been able to back up this story, using evidence both from other solar systems that we have observed during their early life and from investigating the behaviour of our own sun. Powerful modern telescopes have revealed that some apparently young stars, such as Beta Pictoris, are surrounded by disks of cool dust, just as the nebular hypothesis predicts. Meanwhile, many young stars that are too far away for us to see clearly emit an excess of infrared radiation, a fact that could best be explained if they were surrounded by disks of cool material.

We also have good evidence that the rotation of our sun is continuing to slow down, allowing it to continue to shrink. NASA's Parker Solar Probe is currently finding that the solar particles emitted from the sun rotate with it and are released from its magnetic field in just the way that had been proposed. And they are being released much farther out than had previously been thought, some 20 million miles from the sun's core. This provides evidence that verifies how the sun initially shrank into a ball and shows that the sprinkler mechanism is even more effective than had previously been believed. Indeed, the mechanism has been so effective that though the sun still rotates once every twelve hours, this is nowhere near fast enough to keep it at its present size. It is only kept inflated by the pressure caused by the fusion reactions at its core. When its fuel eventually runs out in several billion years' time, the sun will collapse into a white dwarf star with a radius only slightly larger than that of the earth.

The nebular hypothesis is able to explain more than just the birth of our own solar system. The stars that we see in the night sky were also produced in much the same way, and as we are now finding, many if not most of them are also surrounded by systems of planets very like our own. The theory has also been extended to explain phenomena that occur on a much grander scale. As well as suggesting the nebular hypothesis, Immanuel Kant was probably the first person to realise that the shimmering band of light that circles the heavens, the Milky Way, is in fact a huge disk of stars, and that our sun is merely a single star within this huge structure. We now know that our sun is positioned some two-thirds of the way out from the centre of our galaxy, and like the rest of the stars our sun is not still, but rotating about a supermassive black hole at the galaxy's centre. Like our solar system, our galaxy was also created from condensation of a spinning cloud. However, it was formed far earlier, around 13.6 billion years ago, shortly after the big bang, and from a far larger cloud of the gas that was formed during the creation of the universe: hydrogen. The early stars would have been quite different from our own, which has been recycled from the debris produced by the destruction of earlier stars. Kant also correctly surmised that many of the tiny elliptical smudges in the night sky, which are known as nebulae, are also disk-shaped galaxies, which are oriented at an angle to us and that are located at almost unimaginable distances from our tiny home planet. They, too, must have been formed by spin.

So spin really did create both the heavens and the earth. Which leads to the question of what created the clouds of gas that formed the galaxies in the first place, and what caused them to spin. The answer seems to lie right back at the start of the universe, in the big bang. Recent measurements of the background microwave radiation, the echo of the big bang, have shown fine-scale graduations in intensity—an indication that the expansion of the universe was not uniform. Just as a conventional explosion sets up a whole series of eddies in the air that it displaces, so the big bang formed huge swirls in the clouds of gas that it produced. It was these eddies that acted as the nuclei for the formation of the huge range of galaxies, black holes, stars, and planets that make up the known universe. Spin is the very reason our universe is here at all.

CHAPTER 2

How Spin Made the Earth Habitable

When the earth and moon had been formed, some 4.5 billion years ago from a massive interplanetary car crash, they found themselves in the "Goldilocks zone." They were orbiting the sun in a region where its radiation could keep the surface of a planet at balmy temperatures between the freezing and boiling points of water; they were in a zone where liquid water could exist and where life could evolve. However, at that point in time there seemed to be no prospect of life ever emerging. For the earth's interior was so hot that even at its surface the rocks were molten, and it was covered by a dense atmosphere of carbon dioxide 3 million times as dense as today's. This created a massive greenhouse effect, effectively insulating the earth's surface from the cold of space. There seemed to be no way that it would ever cool down. However, within 10 million years the situation was completely different. The surface crust had solidified; tectonic plates were moving across the earth's surface; and most of the carbon dioxide had dissolved into the rock and been dragged down into the earth's mantle as the plates were subducted at plate boundaries. The earth's surface had become much cooler and much of it was now covered in liquid water. The conditions were starting to become perfect for life. The key to this rapid change was the spin of the earth and the orbit of the moon. And surprisingly it was down to forces that also produce a modern phenomenon that locally increases biodiversity, but which is hardly transformational: tides.

The clue that oceanic tides are caused by the action of the moon is that high and low tides coincide with the rise and fall of the moon. The clue that the sun is also involved is that the strength of the tides varies with the moon's phases. However, it took the genius of Isaac Newton to explain how this miracle was powered by the moon's gravity. It is not surprising that people found this hard to believe at first, because the effect of the moon's gravity on earth is tiny. Of course, just as the gravity of the earth attracts the moon to keep it in orbit, so the moon's gravity attracts the earth, but the acceleration it causes is just three-millionths of the earth's gravity. However, because the water on the side facing the moon is closer, it is attracted to it slightly more than the planet itself, around a tenth of a millionth of earth's gravity, while the seawater on the far side will be attracted about a tenth of a millionth of earth's gravity less. It does not sound like much, but since this acceleration acts perpetually, this is enough to move the water large distances; the seas have long since flowed towards and away from the moon and formed bulges on both the near and far sides of the earth. And as the earth spins past the moon, the bulges wash around the surface of the earth, forming the tides: high tides when the moon is high in the sky or below the horizon, and low tides when the moon is rising or setting.

The sun has a similar, if smaller, effect on the world's seas. The sun's gravitational pull on the earth is around 180 times greater than that of the moon, but because the sun is so much farther away, the difference between its gravity on the near and far sides of the earth is only around a third that of the moon. This would cause the water to bulge to a correspondingly smaller extent when the sun is overhead (at midday) and when it is on the far side of the earth (midnight). The two sets of tides, lunar and solar, interact to form a consistent daily and monthly pattern. We get two high tides every twenty-five hours as the earth spins around to the same position relative to the moon, but the strength of the tides varies over a fourteen-day period. They are strongest, so-called spring tides, when the lunar and solar tides coincide, at full and new moons, and weakest, so-called neap tides, about half the size of spring tides, when they oppose each other, during half moons. The average height of the ocean's tides is actually tiny—ranging from 12 inches (30 centimetres) for neap tides to 36 inches (90 centimetres) for spring tides, but as they reach the shore and the water depth falls, the height of the tides is magnified, just like the waves produced by tsunamis, and the effect can be magnified further if water moves in and out of a funnel-shaped bay. In the Bay of Fundy, in Eastern Canada, for instance, the distance between high and low tides can be up to 55 feet (17 metres).

As anyone who has lived, or had holidays by the seaside, knows, the effect of tides can be dramatic. The rise and fall of water around the world's shores creates the miraculous phenomenon of the regular covering and uncovering of sandy beaches, mangrove swamps, coral reefs, salt marshes, and rocky shores. Intertidal habitats contain a vast variety of seaweeds, corals, sea grasses, and mangroves, which in turn provide food and shelter for an even greater variety of animals. The spin of the earth has given the world the amazing intertidal ecosystems that hold over half of the sea's biodiversity.

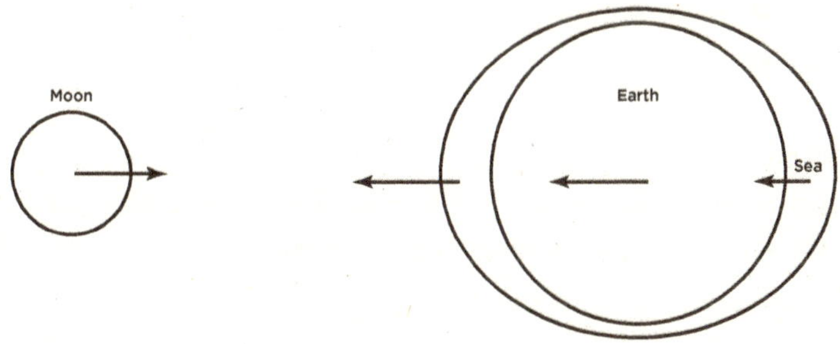

The explanation of the diurnal tides. Because gravity is stronger on the side facing the moon, and lower on the side facing away, the water bulges out on both sides, and the bulges are swept around the earth as it spins.

However, the benefits the tides confer to intertidal creatures and holiday-makers are just a minor aspect of the effects that they have had on our planet. They also have had much larger global impacts.

Tidal forces have changed and are continuing to change the speed at which the earth spins and the distance at which the moon rotates about the earth. The energy needed to sweep the seas across the globe comes from the kinetic energy of the spinning earth. As the oceans sweep westward around the earth their motion is resisted by friction with the seafloor, so one of the main effects of the tides is to slow down the eastward rotation of the earth. Days are consequently getting 2.3 milliseconds longer every century. And just as the moon affects the tides, so the tides affect the moon. Because the centre of the bulges of water are slightly in front of the moon, gravity is gradually acting to pull it forward, increasing its angular momentum and causing it to be slung 1.4 inches (3.5 centimetres) farther away from the earth every year. In the future the earth will slow down still further, the moon will be slung farther away, and given enough time the lunar tides will get weaker, until finally solar tides dominate and high tides will occur exactly twice a day. And meanwhile the earth will continue to slow down so that the days get longer. However, we need not worry too much about this. Long before the earth's spin slows to a halt, some 1.5 billion years in the future, the sun will have expanded and burned away the oceans, dramatically reducing the rate at which the earth slows down. And by 4 billion years' time the sun will have become a red giant and engulfed the earth in a flaming mass of gas.

But just as in the future the earth will spin more slowly and the moon will be farther away, in the past the reverse must have been true; the earth must have spun faster and the moon would have been closer. In fact, soon after the birth of our planet and its consort, the moon was only 15,000 miles (24,000 kilometres) away, sixteen times closer than today, and it would have orbited the earth once every ten hours or so, while the earth itself spun around once every six hours. The tidal forces would have been far greater, large enough to squeeze the molten earth into a lemon shape, churning the mantle so that it released more heat, like porridge being stirred. The huge friction caused by the internal movements of magma caused the earth to slow rapidly, and the moon to be slung farther away, so that the tidal forces quickly fell. The change was so rapid that within a few million years the surface of the earth started to solidify. The solid rock was finally able to absorb some of the carbon dioxide in the atmosphere and draw it down into the earth as the tectonic plates subducted into the mantle. The earth cooled increasingly rapidly, allowing water to condense to form the oceans just 10 million years after the cosmic collision.

By this stage, tidal forces had slowed the earth's spin to once every ten hours and the moon had been flung out to 80 percent of its present distance. The first oceanic tides on earth were just 40 percent more powerful than today's, and the forces slowing the earth were little greater than they are today. The days continued to get longer, but only gradually. By 3.5 billion years ago, when life was first emerging, days were around twelve hours long. By 1.4 billion years ago, when the first eukaryotic cells were appearing, day length was around eighteen hours; by 500 million years ago when the first multicellular organisms were emerging onto land, days were around twenty hours, fifty minutes long; and by the end of the reign of the dinosaurs, just under 70 million years ago, days were twenty-three hours and thirty minutes long, just half an hour shorter than today, and there were 372 days in a year.

Tidal forces have had one further effect that is visible even to the naked eye when we look up at the night sky. When it was created, the moon was spinning rapidly, just like the earth and rotating in the same direction. However, since the earth has a mass that is around eighty-one times that of the moon, it produced much greater tidal forces on the moon than the moon did on the earth. The earth's gravity therefore caused the early moon to bulge much more than the earth, deforming into a lemon-shaped body. And as the moon spun around and changed shape, internal friction rapidly slowed its spin. The effect was so great that the tidal forces soon stopped the moon from rotating at all, relative to the earth. Today the moon is tidally locked to the earth. It spins just once every twenty-eight days, effectively still relative to the earth, so that the same side of the moon is always pointing towards us. We can all gaze at its rough mountains and smooth seas, and try to make out the "man in the moon," but only a few astronauts have ever set eyes on the far side of the moon.

CHAPTER 3

How Spin Stabilises the Earth

At the end of the last chapter we left the earth orbiting the sun at its present distance of around 93 million miles (150 million kilometres) and spinning around its axis at an angle of 23.5 degrees to its orbit; and we left the moon orbiting the earth, along a plane that is oriented at 5 degrees from the earth's orbit. These angles have remained remarkably constant throughout the earth's existence. Over their 4.5-billion-year history, the earth's spin has slowed down a bit and the moon has been slung a bit farther away, but nothing much else has altered. One might have expected that, in over 4 billion years, the earth might have flipped over a few times, like Dzhanibekov's wing nut; or other planets, especially our nearest large neighbour, Venus, might have perturbed its course and caused it to wobble or tilt. After all, football balls and plates are all too prone to wobble about or turn as they fly through the air. The fact that the earth has been stable for so long has been crucial for the emergence and continued existence of life on this planet. We have been extremely fortunate that we live on a planet that rotates in a near circular orbit, and at just the right distance from the sun so that its average surface temperature lies between the freezing point and boiling point of water. But that would be no use for life if the earth tumbled erratically; any point on its surface might alternate between periods of such extreme heat and cold that no organism could survive. As we shall see, part of the reason that the earth is stable is down to its spin; the rest is due to our relationship with our neighbour, the moon, and the way our two orbs rotate about each other.

We usually assume that the earth is more or less perfectly spherical, because gravity will have long since drawn the material as close to its centre as possible. Certainly, the pressure at the centre of the earth due to all the rock pressing down from above is more than enough to compress it into a sphere; it is an astonishing 364 gigapascals, almost 4 million atmospheres! However, because the earth is spinning, as we saw in chapter 1, each part of it is also accelerating inward, centripetally towards its axis. According to Newton's third law, therefore, this will apply an *outward* centrifugal force on the material. The earth is consequently stretched outward by its own spin. It's not by a large amount. Since the earth only spins once a day, the centrifugal force on an object standing on the equator is a mere 0.3 percent of the force of gravity at that point. Consequently, unlike a pizza base that flattens into a thin crust when it is tossed, spinning into the air, or a molten glass vessel that bulges outward when spun around by a glassblower, the effect is not noticeable, even from space. But it is measurable and important. The spin has deformed the earth into an oblate ellipsoid, flattened at the top and bottom and bulging at the sides, like a mandarin orange, so that its diameter across the equator is 30 miles (48 kilometres) greater than it is from pole to pole.

This bulge, though small, effectively stabilises the earth's axis. If the earth tilted, the centrifugal forces acting on the bulges as it spins around would act to return it to its original orientation. Just like a plate spinning on a stick, or a Frisbee spinning through the air, the spin of the earth maintains its orientation. The effect is weak, however, because of the small size of the bulge. It has been estimated that if the earth was spinning on its own through space, the gravitational pull of the other planets in the solar system, particularly Venus, would have tilted its axis by a full 90 degrees. Fortunately, the earth is kept in its orientation even more firmly because it is held in the grip of a much more stable system. Because the earth and moon rotate about each other at such a great distance, they, too, are stabilised by the centrifugal forces that would generate a much larger restoring torque if they tilted. They rotate stably around each other, just like the weights at the end of a majorette's baton as it is twirled around.

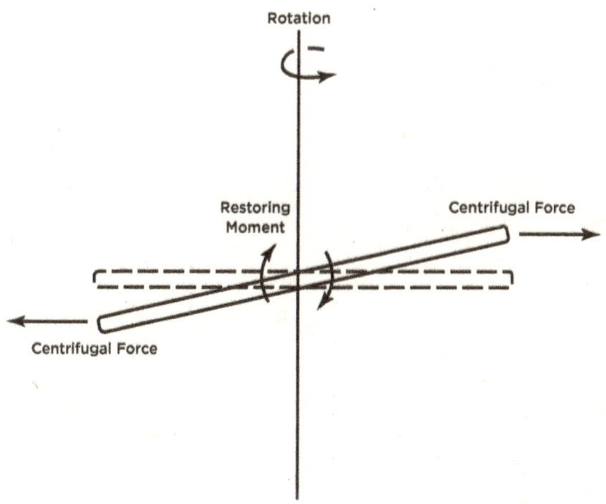

Stability of a spinning plate. If the axis tilts, the centrifugal force on the raised and lowered sides will produce a restoring moment tending to return it to its original orientation.

The bulge of the earth does cause one potential problem, however. Since the moon orbits the earth at a lower angle than the earth's axis, and since its gravitational pull on the near side will be greater than that on its far side (just as they are on the world's oceans), it will exert a small torque causing the tilt of the earth's axis to decrease. The sun has the same, if somewhat smaller, destabilising effect, just as it has a smaller effect on the earth's tides and for much the same reason; the sun's pull is greater than that of the moon, but being so far away the sun's gravity on the near and far sides of the earth are more equal. As a consequence of these torques you might expect the earth's axis to have long ago been pulled upright so that it spins parallel to its orbit, banishing the seasons. But since the earth is spinning it does not behave like a simple stationary object, but like a spinning top or gyroscope that is leaning away from its support and being pulled over by the force of gravity. And, despite what Eric Laithwaite said in his Christmas lectures back in 1974, the behaviour of a gyroscope is really quite easy to understand, and can be readily explained using the known laws of physics.

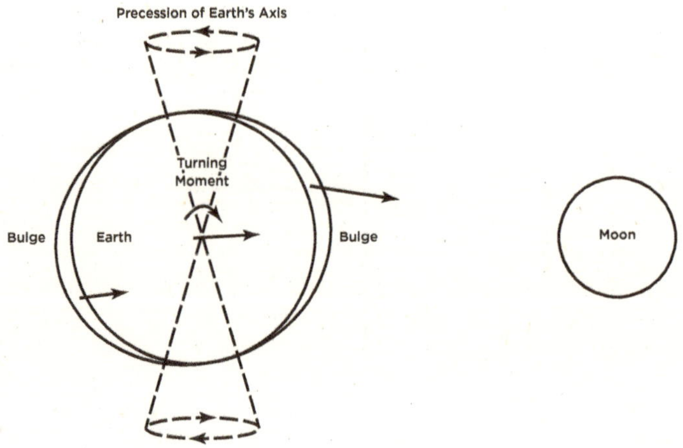

The precession of the earth. The difference in gravitational attraction to the moon on the near and far sides of the earth's bulge produces a turning moment that should return its axis to the orbital plane. Instead, however, it wobbles or precesses once every twenty-six thousand years (*small arrows*).

What initially happens when you release a spinning gyroscope that has one end resting on a support and is leaning over is exactly what you would expect to happen if it was not spinning: it starts to fall over, tilting farther under its own weight. But as it tilts, something interesting happens to the material spinning around its rim. The parts of the rim on the side moving downward are accelerated *inward* because of the downward movement of its axis (see the diagram opposite), while the parts of the rim on the side moving upward are accelerated *outward*. The reaction to these accelerations automatically sets up a torque that pushes the gyroscope sideways, towards the side of the rim that is moving upward. The gyroscope will start to rotate around a vertical axis through its base, a motion known to physicists as precession. And as the gyroscope moves sideways as well as downward, the reaction force, which is at right angles to the motion of the gyroscope's axis, will turn upward as well as sideways. The gyroscope will move along a cycloidal path, like the movement of a point on the rim of a rotating wheel, and because the upward moment will soon exceed the effect of the weight, the gyroscope will "bounce" upward again and slow down. In an ideal world the gyroscope would continue travelling around the axis in a series of cycloids. However, because of friction, the oscillations will gradually die away and the gyroscope will eventually settle into a steady sideways motion, at which point the upward force produced by the precession equals the downward force of gravity. The gyroscope will move around with its axis at a slightly lower angle than where it started.

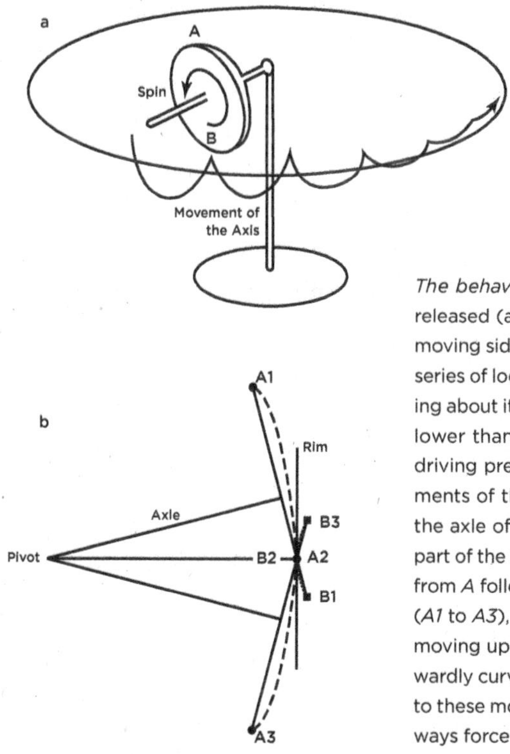

The behaviour of a gyroscope. When released (a), it first starts to fall before moving sideways and upward again in a series of loops, before precessing, rotating about its base around a vertical axis, lower than when it started. The force driving precession is caused by movements of the rim of the gyroscope. As the axle of the gyroscope falls (b), the part of the rim that is moving downward from *A* follows an inwardly curved path (*A1* to *A3*), whereas the part of the rim moving upward from *B* follows an outwardly curved path (*B1* to *B3*). Reaction to these movements produces the sideways force that starts precession.

Once we understand the basic mechanics, it is also easy to see what's going on when we wind up a spinning top and set it down vertically on its point. At the start, when it is spinning fastest, it will drop only a tiny amount because the torques set up by the spin will be large, and it will precess only slowly. But as friction slows down the spin, for a given rate of precession, the gyroscope will produce less and less restoring torque; it will start to drop again until its precession speeds up enough so that the restoring moment once again balances the weight of the top. Gradually the axis of the top will tilt farther and it will rotate in wider and wider circles and precess faster and faster. Its movement will get wilder and wilder until eventually the rim hits the floor and the top careers around the room.

Toy gyroscopes and tops spin rapidly, so they typically precess about once every second. In the case of the earth, however, the precession caused by the action of the sun and moon is very slow. Partly this is because the earth spins so slowly, only once a day. And the restoring torques of the moon and sun effectively act only half of the time because the earth's axis is rarely oriented parallel to the line between the celestial bodies. But the main reason is because the bulge of the earth is very small compared to its diameter; the difference in the gravitational accelerations on the bulges on the near and far sides are therefore tiny compared with the huge mass and size of the earth. As a consequence, the earth precesses only once every twenty-six thousand years, the so-called "precession of the equinoxes" that has been observed by astronomers down the ages. This precession is a main contributor to what are known as the Milankovitch cycles. Along with drift in the elliptical orbit of the earth, which varies every 110,000 years, and wobbles in its axis, caused by interference with other planets, which varies every 41,000 years, the earth's precession controls the amount of heat the earth receives from the sun. These cycles consequently drive the onset and duration of ice ages.

The earth's axis also undergoes a couple of minor movements over much shorter time scales. The moon's orbit precesses every 18.6 years, for instance, which causes the earth to wobble with the same frequency, but only by about 160 feet (49 metres). And the earth itself wobbles, like a spinning plate, every 433 days, buffeted by storms on its surface, but this is even less worth worrying about as the movement is a mere 30 feet (9 metres)! For all intents and purposes, the earth's surface is stable. And as long as it continues to rotate, the earth, distorted and stabilised by its spin, and kept even more stable by being shepherded by our large companion moon, will remain safely orbiting for many hundreds of millions of years into the future. There is no danger of a Dzhanibekov flip.

CHAPTER 4

How Spin Shields the Earth

From the pioneering science fiction tale, Johannes Kepler's *Somnium*, which describes an imaginary journey to the moon, novelists and scientists alike seem to have been obsessed with the idea of space travel. In particular they seem to love the prospect of humans visiting and even living on the moon and on other planets, especially Mars. However, quite apart from the logistical difficulties and huge expense involved, there is one major barrier that might make colonising space impractical: its radiation environment. Throughout our solar system, an astronaut would be assailed from all sides by deadly ionising rays. As we saw in chapter 1, the sun emits a stream of charged particles, mostly alpha and beta particles (helium nuclei and electrons respectively), streams that wax and wane along with the appearance and disappearance of sunspots over the eleven-year solar cycle, and which can flare up over periods of a few days into intense solar storms. The astronauts would also be rained on by a slower stream of cosmic rays from outside the solar system, rays that consist of charged nuclei of helium, oxygen, and iron. Spaceships have to be shielded from this radiation using multiple layers of aluminum, Kevlar, and epoxy resin, interspersed with air spaces that can stop or slow down most of the particles. And space walks are limited to a few hours at most. The astronauts on America's Apollo moon missions were also lucky that their voyages never coincided with solar flares; if they had, their missions would have had to be aborted and the astronauts returned as soon as possible to earth so they could be treated for the effects of severe radiation sickness.

You might expect, therefore, that back on earth we, too, should have problems with cosmic rays. It would not be such a problem for life in earth's oceans, because the radiation would be absorbed by the first few feet of water. However, once organisms emerged from the water and became terrestrial they would have had to run the gauntlet of the full range of ionising radiation, which would have destroyed the DNA in their cells. And there is a good chance that the radiation would have long ago stripped the earth of its atmosphere, just as it has done on Mars. We are fortunate, then, that here on earth we are protected from these dangers by our planet's magnetic field; the earth acts like a giant bar magnet producing a magnetic field, which runs in a huge ring from the magnetic South Pole to the magnetic North Pole and which projects far out into space. The field protects the earth by deflecting most of the incoming charged particles. And the few particles that do penetrate into the magnetic field are caught and corralled, circling along the field lines, before becoming trapped in two regions of space, the Van Allen belts that circle around the earth at heights of between 0.2 and 2 times its diameter and between 3 and 10 times its diameter, well above the atmosphere. They are held far above the region at which manned spaceships such as the Space Station orbit the earth, explaining why the astronauts in these vehicles can stay in space for so long. The earth's magnetic field keeps us safe from all but a very few cosmic rays. And like so much else about our planet, we have spin to thank for its existence.

But it's not the earth's spin alone that causes the magnetism. The reason we are protected results ultimately from events that occurred at the birth of our planet. As we saw in chapter 1, when the earth condensed from cosmic dust 4.5 billion years ago, the gravitational energy was converted to heat. This melted the rock, allowing the earth to condense into a near spherical shape, with the heaviest metals—including nickel and iron—dropping down into the core that stretches out halfway to the earth's surface. You might have expected the earth to have cooled down enough since then to allow the core to solidify. Indeed, in the late nineteenth century the British physicist William Thomson, Lord Kelvin, calculated that this would take just a few tens of millions of years. Fortunately for us, however, the centre of the earth continues to be heated by an energy source unknown to Lord Kelvin, the radioactive decay of heavy atomic elements. Consequently, though the earth's inner core, extending one-fifth of the way from the earth's centre, has solidified, the outer core remains a liquid and the radioactive decay powers huge convection currents within it. As the liquid iron and nickel heat up near the centre, the metals expand and rise, displacing cooler metal farther out, which sinks back down to the inner core. In a stationary planet the metal movements would form huge convection currents, just like the ones you see when water is heated in a saucepan, but oriented radially. These plumes would be so chaotic that over the whole planet the movements would cancel one another out and there would be no way they could be harnessed to have any net magnetic effect. However, the spin of the earth changes everything because of the response of fluids when they are in a body of fluid that is rotating.

To see what should happen in a rotating fluid, it is best to examine a series of simple experiments that the British physicist G. I. Taylor performed in 1914, on beakers of water that he set spinning. If you rotate a liquid-filled vessel on a turntable, the whole body of liquid soon starts to spin with it, and the surface rises away from the centre in a parabola until the extra pressure produced by the higher column of liquid farther out cancels out the centrifugal force on the water. Just like the sea lying on the ellipsoidal surface of the earth, the system is stable. The interesting thing is what happened when Taylor sought to alter the balance by perturbing the water within the vessel. For instance, if he dropped a drop of dye into the water it did not spread out gradually around the whole vessel as you would expect. Instead, the heavy dye particles moved downward, but if they started moving sideways, they were quickly diverted into moving in a narrow circle. The result was the formation of what are known as Taylor columns. This behaviour is just one example of what happens when objects move about in a system that is already rotating, behaviour that was first comprehensively analyzed in 1835 by the French mathematician Gaspard-Gustave de Coriolis.

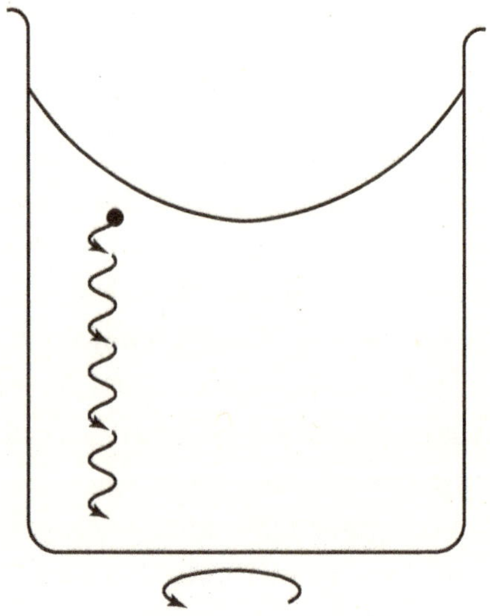

Taylor columns in a spinning beaker of water. If you drip dye into the water, the heavy dye particles do not spread out evenly as they would in still water but are diverted into a helical path as shown and circle directly downward.

We have just seen that the inward force caused by the extra pressure farther out in the water column will exactly balance the outward centrifugal force caused by the rotation of the water. However, if the water is disturbed so that the dye particle moves at right angles to the axis of rotation, this will no longer be the case. If it moves in the same direction as the rotation, its absolute velocity will increase, and so will the centrifugal force on it. It will now be greater than the inward pressure and this will produce an extra force that will cause it to be deflected outward. In contrast, if the dye particle moves in the opposite direction to the rotation, its absolute velocity will *decrease* and so will the centrifugal force on it; it will be deflected inward.

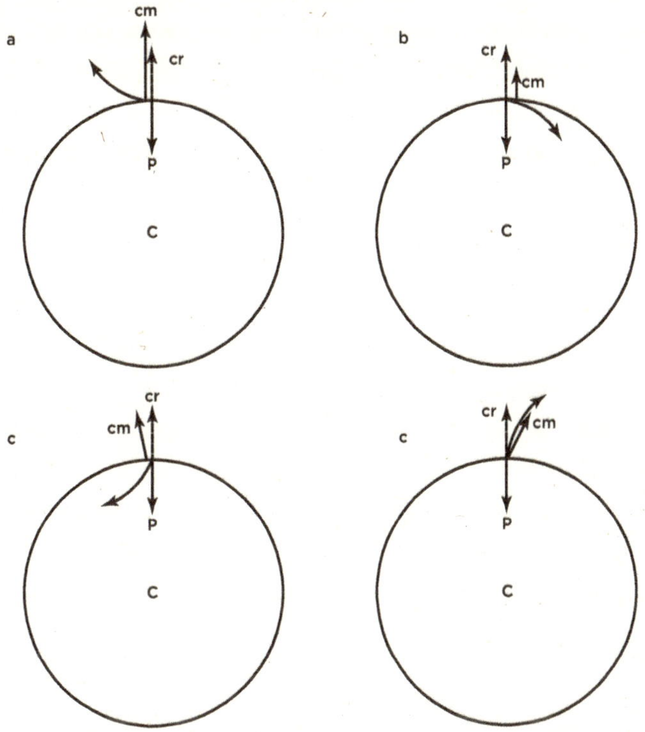

The effect of the spin of a body of water on the motion of a dye particle. When the dye particle is moving counterclockwise in a circle along with the water, the inward pressure, *P*, due to the rising meniscus, exactly counteracts the outward centrifugal force (*cr*) on it. However, if the body moves in the same direction as the rotation (a), the centrifugal force (*cm*) increases, becoming greater than the pressure, *P*, and the body is forced outward, to the right. If the body moves in the opposite direction (b), the centrifugal force falls and the body is forced inward, again to the right. If the body moves inward (c), its path is converted into an inward spiral and the centrifugal force displaces it forward, to the right. And if the body moves outward, the path is converted into an outward spiral and the centrifugal force displaces it backward, again to the right.

If the dye particle moves towards the axis of rotation, its path will be changed from being a circular motion to being an inward spiral. The centrifugal force on it will now act at right angles to its path, outward and forward. Finally, if it moves outward, its path will be changed from being a circular motion to being an outward spiral. The centrifugal force on it will now act at right angles to its path, outward and backwards. If the beaker is rotating clockwise, the dye particle will always be deflected to the right, counterclockwise, while if the rotation is counterclockwise, the water will always be deflected to the left, clockwise. Since the motion affects the magnitude and direction of the centrifugal force on the particle, Coriolis described the deflecting force as the *force centrifuge composée*, or "compound centrifugal force." We know it today as the Coriolis force, an apparent force that acts at right angles to fluids moving in a rotating reference frame, a force that can deflect but not accelerate the fluid. Taylor columns are an example of the effect Coriolis forces have not only on particles but also on fluids: they divert the motion of both particles and fluids into circles and reduce mixing. The same is true of fluids that are subjected to larger perturbations. When Taylor heated his spinning beakers of water from below, the water did not rise and fall in huge convection plumes as they do in still water, but in many narrow vertical columns.

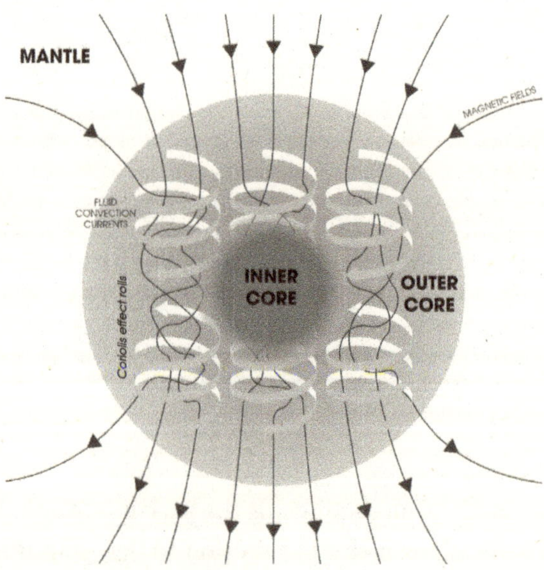

A diagram showing the mechanism that produces the earth's magnetic field. Convection currents in the liquid metal outer core are diverted into rotating columns by Coriolis forces and power a self-sustaining dynamo that produces the magnetic field.

Now that we can understand the behaviour of fluids moving in a rotating system, it is clear that Coriolis forces will also divert the convection plumes of metals in the earth's outer core. Rather than rising and falling in radial columns, Coriolis forces will divert them into a series of spirals oriented north to south. It is this motion that produces the earth's magnetic field.

How it produces a magnetic field is in turn determined by the laws of electromagnetism, which were worked out in the middle of the nineteenth century. As the great British physicist Michael Faraday showed, if you spin a metal disk through a magnetic field, this induces a small electrical current in the disk, a current that is oriented at right angles to both the field and the movement. Likewise, running an electric current through a moving cylinder produces a magnetic field. The efficiency of both devices can be improved by using not solid metal but insulated wire-wound solenoids, the same shape as the spiral currents in the earth's outer core. Another major advance was made by the Hungarian Anyos Jedlik, who in 1856 developed a dynamo that used the principle of dynamo self-excitation; both its stationary and revolving coils were electromagnets, and their interaction produced the magnetic field that in turn produced the electricity. No permanent magnet was needed. In the earth's core these two processes also interact in this way. If the spiralling metal moved through a small initial magnetic field, this would produce an electric current that would in turn produce a larger magnetic field. The process would soon be self-sustaining, and the earth would act as a self-exciting dynamo, like the ones used in so many electric generators today. So it is the spiral motion of the iron and nickel that produces electricity within the earth's core, and consequently the magnetic field that surrounds the planet.

Relying, as it does, on fairly chaotic convection processes to power it, the earth's magnetic field is not constant; it varies slightly both in strength and direction, what is known to scientists as secular variation, explaining why magnetic north seems to wander randomly across the Arctic. And the field can even collapse every few hundred thousand years and reverse its direction. This leaves

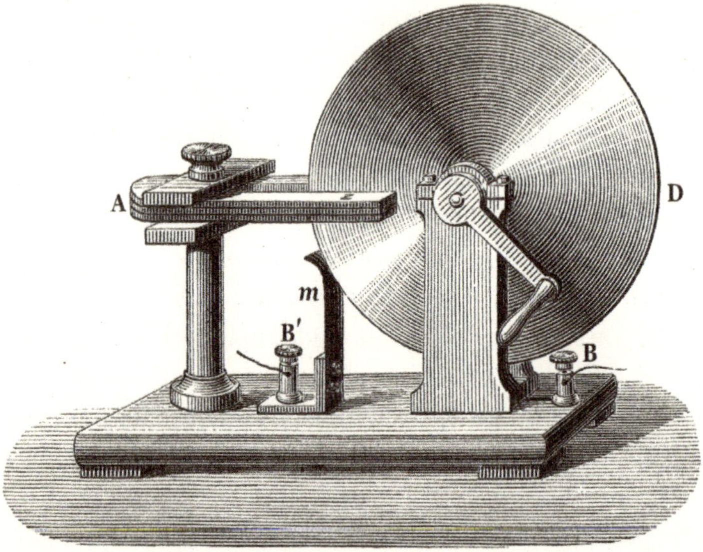

Faraday's experiment showing that if a metal disk (*D*) is rotated within the arms of a permanent magnet (*A*), electricity is produced.

the earth temporarily exposed to the solar wind, and produces the changing magnetic bands in rocks that helped geophysicists uncover the secrets of continental drift and plate tectonics. We are fortunate that the last complete reversal of the earth's magnetic field took place some 780,000 years ago, because a short-lived event some 42,000 years ago, when the earth's magnetic field lost around 95 percent of its strength, shows how damaging these events can be. It coincided with ecological disruption that may have helped to kill off many large mammals and the last remaining members of our closest cousins, the Neanderthals. Ever since, our species has not had to cope with that challenge and we have been protected from the solar wind by the earth's spin.

And spin has left us one further aesthetic gift. During intense solar storms, some charged particles do manage to escape the Van Allen belts and spiral along the earth's magnetic field lines down into the atmosphere in the polar regions. Fortunately, they can do little damage in these frozen wastes because there are few terrestrial organisms around to harm. In the polar sky they collide with atoms in the upper atmosphere, exciting the electrons in their outer shell. It is when these electrons drop down back into their original state that they release the energy, in the form of light, that produces the beautiful auroras of the two polar regions, the northern and southern lights.

CHAPTER 5

How Spin Controls the Earth's Climate and Weather

The problems with cosmic rays mean that humans would always have to be shielded if we were to colonise Mars or the moon. But even if we did manage this, it seems to me that life on these bare, rocky bodies would be hardly worth living. We would surely miss the dappled sunlight in sylvan glades, the nodding heads of flowers in the meadows, and the fields of rippling wheat. We would surely long for the ever-changing nature of the planet, driven by the march of the seasons: the freshness of spring, the heat of summer, the mists and mellow fruitfulness of autumn, and even the snows of winter. And we would miss the short-term unpredictability of the weather: the breeze on our cheeks, the sunshine after a rain shower, and the miraculous arc of the rainbow. We may enjoy holidaying in places where we can rely on the weather—the dry heat of the desert, the cold of the mountains, and the steamy heat of the tropical rainforest—but most people prefer to live in temperate regions. Even the perennial sunshine of Los Angeles can pall. So, to me at least, living on another planet would be like being confined in a stone prison.

We are fortunate indeed that spin has given us one final blessing. It controls the convection currents in our atmosphere, just as it does those in the earth's core, moderating temperatures, reducing wind speeds, and regulating rainfall. In doing so it creates a range of benign climatic conditions in which a wide diversity of plants and animals can thrive. It makes the world a marvelous place in which to live.

The first consequence of the earth's spin is, of course, that the surface of our planet is alternately heated by the action of the sun during the day and allowed to cool when on the far side of the sun, at night. You would think that this might produce large temperature fluctuations that would power strong winds around the world. Fortunately, however, temperature changes are kept to a minimum by two things. First, the earth spins so fast that days and nights are too short—an average of just twelve hours—for the earth to heat up and cool down by too much. Second, the greenhouse effect, caused by the water vapour and carbon dioxide in the earth's atmosphere, reduces the escape of infrared radiation at night, insulating the ground and reducing rates of energy loss. Consequently, differences between day and night temperatures are small, being a mere 9 degrees Fahrenheit (5 degrees Celsius) in the cloud-blanketed tropics, and at most 75 degrees Fahrenheit (40 degrees Celsius) in cloudless deserts. This limits the effects that periodic solar heating might otherwise have on atmospheric pressure, and so on the wind. The solar tides that it produces, and which travel westward around the earth, following the sun, are very weak. They produce only tiny fluctuations in pressure in the lower atmosphere and have little effect on our weather, though at high altitude, in the mesosphere, 30 to 60 miles (50 to 100 kilometres) up, they can drive wind speeds of up to 125 miles per hour (200 kilometres per hour).

By far the greatest effect of solar heating is to produce convection currents in the earth's lower atmosphere, currents that are driven by the difference in heating between equatorial and polar regions. If the spin of the earth had no effect on the atmosphere, solar heating at the equator would drive two huge convection currents around the world. The heat of the sun in the central tropics would warm the air and evaporate water vapour from the seas and from the vegetation that covers the land. This would lower the air density, both because the warmer air molecules would be spread farther apart, and because more of them would be the light H_20 molecules (molecular weight 18) rather than the heavier N_2 or O_2 molecules (molecular weight 28 and 32 respectively). The lighter, wetter air would rise and cool, releasing much of its water vapour in the form of clouds and rain, before moving at high altitude towards the poles, where it would finally cool and fall back down to the earth. The high pressure this set up at the poles would then drive the air back towards the equator. This system would appear to be benign, but as the winds would travel uninterrupted for thousands of miles, they would build up to form perpetual hurricanes: northerlies in the northern hemisphere, and southerlies in the southern hemisphere. It would make both hemispheres uninhabitable.

Fortunately, the spin of the earth does deflect the winds that move across its face, just as it deflects the movements of the metals in the earth's core. In the case of the atmosphere, the situation is rather more complicated, however, because the strength of the Coriolis forces depends on the latitude. Near the poles, any wind blowing across the ground will move at right angles to the earth's rotational axis, so it will be deflected strongly sideways—in the northern hemisphere winds will be diverted to the right; in the southern hemisphere to the left. In contrast, at the equator, winds travelling north or south will be travelling parallel to the earth's axis and will not be affected at all by the earth's spin; and winds travelling east or west will only be deflected upward or downward. The strength of Coriolis deflections will change progressively with latitude; the lower the latitude the smaller the effect of Coriolis forces.

A major consequence of the earth's spin is that the convection currents get diverted so that rather than being surrounded by a single huge convection cell, each hemisphere is covered by three smaller ones. At the equator the warm wet air rises before starting to travel directly to the poles. However, as it does so, Coriolis forces start to deflect it eastward until the poleward motion is stopped at a latitude of around 28 degrees. There it piles up, forming areas of high pressure, which push air downward. This air warms up as it falls, evaporating the water vapour, so the region is characterised by cloudless skies, low rainfall, and desert conditions. Finally, the high pressure drives the air back to the equator at low altitudes. And in turn these winds are deflected westward by Coriolis forces, forming the trade winds of the outer tropical regions, before coming back to rest in the doldrums in the central tropics.

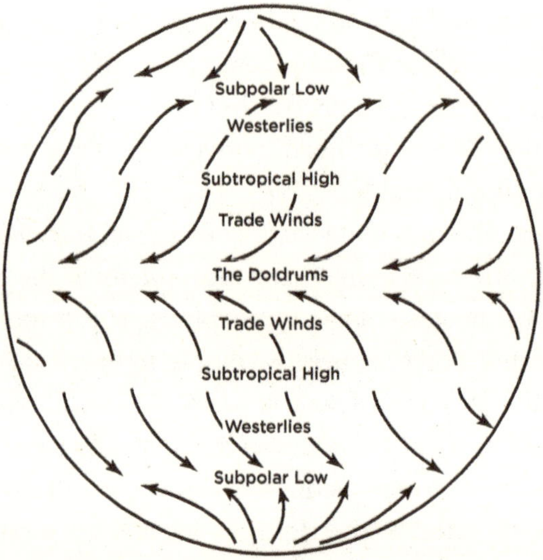

The global pattern of winds and atmospheric pressure as first shown by the nineteenth-century meteorologist William Ferrel.

The high-pressure areas at 28 degrees also drive some air towards the poles, and because the strength of the Coriolis force increases with latitude, the winds are quickly deflected to the east, forming the predominant westerlies of the temperate regions. The continued deflection finally stops the poleward motion of air at latitudes 60 to 70 degrees. This air meets winds travelling down from the poles and rises before travelling at high altitudes back to the tropics or back to the poles. Rather than having one convection cell in each hemisphere, therefore, the earth's rotation results in the formation of three: the first between the equator and 28 degrees (now known as Hadley cells); the second between 28 degrees and 60 degrees (now known as Ferrel cells); and less-defined polar cells between 60 degrees latitude and the poles.

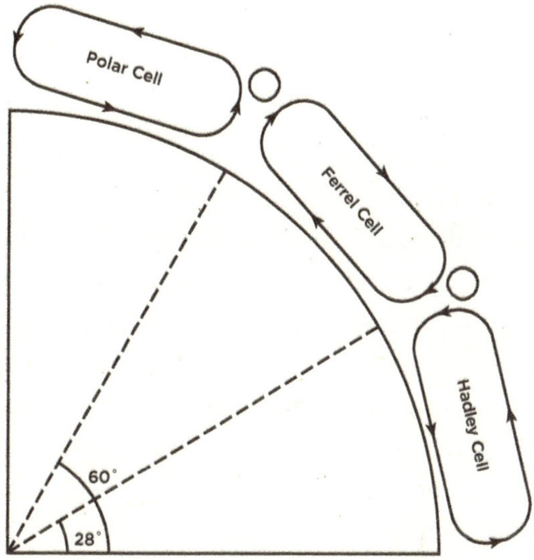

The convection driven movements of air in the atmosphere. See the text for the explanation. The circles are the more recently discovered jet streams.

The existence of the three sets of convection cells readily explains why the earth's climate varies so systematically with latitude. They are responsible for the stifling humidity and daily downpours of the central tropics that enable the tropical rainforest to flourish; for the perfect climate of paradise islands in the outer tropics—the Caribbean, Seychelles, Hawaii, and the South Seas—which are bathed in gentle trade winds; for the presence of the high-pressure desert regions of the subtropics like the Sahara and Sonoran Deserts; for the seasonal windy climate of temperate areas; and for the dry polar steppes. And the way that the winds are deflected by Coriolis forces is responsible for the way that climate differs between land that is located on the west and east coasts of the major continents. Since in the subtropics the trade winds blow from the east, the eastern coasts of the major continents—areas such as Florida, the Caribbean, Brazil, the Philippines, and Queensland encounter water-laden winds from the sea and have high rainfall. In contrast, western Mexico and Chile, in the rain shadow of the Rocky Mountains and the Andes, are notoriously dry and are covered by deserts, including the driest place on earth, the Atacama Desert. In contrast, in temperate areas it is the westward-facing coasts that are wettest, including Washington State and British Columbia in North America, and the whole of Northern Europe, while the eastward-facing coasts of Korea and New England are cooler and drier.

A further complication is that climate is also affected by the inclination of the earth's axis, which leans at 23.5 degrees. The change in orientation of the earth relative to the sun as it orbits drives the cells to move north and south every year, creating the annual changes in temperature and precipitation, which cause seasonality. In the outer tropics, for instance, the cloudy, wet conditions powered by having the sun overhead only occur for part of the year, producing wet and dry seasons. In the Indian Ocean they are responsible for the summer monsoon. And seasonal variations are also responsible for the creation of those most benign climates, the Mediterranean ones. These climates are found around 30 to 40 degrees latitude along the west coasts of Europe, California, Chile, the Cape region of South Africa, and Southwest Australia. In winter months these regions are located within the temperate Ferrel cells and are bathed in wet westerly winds that drive plant growth. In contrast, in summer the earth's tilt brings them back into the intertropical high-pressure region, producing the perpetual sunshine that tourists from colder climates love. This perfect Mediterranean climate enables people to live in the open air almost all year-round, enjoy the beauty of the spring flowers, tuck into the wide variety of fruit, vegetables, and olives that thrive in these regions, and drink the wine from the world's best vineyards.

And the winds affect the temperature as well as the rainfall along the coasts of temperate regions. In the Atlantic, the winds form a near circular pattern, blowing southwest in the subtropics, north along the

Eastern Seaboard of the United States, and northeast in the North Atlantic. The wind reaching Northern Europe therefore comes from farther south and so is relatively warm. These winds also drive the surface waters of the Atlantic to form a huge circular ocean current or gyre, the Gulf Stream, which bathes the coast of Europe in warm waters. It is this that has made the European climate so benign that people are able to farm almost as far north as the Arctic Circle. It was a huge shock for the early European settlers of the United States and Canada when they encountered the Arctic winters and snows along the East Coast. Their failure to predict just how long and difficult winters in the new land would be almost resulted in failure of the first colonies.

As well as explaining the *global* patterns of air movement, the Coriolis forces also explain the *local* weather patterns that meteorologists have long been describing, and attempting to understand, in the temperate regions—Europe and North America—where the vast majority of them have always lived. Here, rather than moving directly from areas of high pressure to areas of low pressure, the winds seem to move at right angles! If you have your back to the wind, high pressure would be on your right and low pressure on your left. The wind swirls in circles counterclockwise around areas of low pressure in the northern hemisphere and clockwise in the southern. Meanwhile, air moving away from a region of high pressure is deflected clockwise in the northern hemisphere and counterclockwise in the southern. The reason is that in temperate regions the strong Coriolis forces deflect air travelling over the earth to the right in the northern hemisphere and to the left in the southern hemisphere. So rather than travelling straight towards a low-pressure region where the sun's heat has created a local area of low pressure, the air moves in a circle. In what is known to meteorologists as geostrophic flow, the forces due to air pressure and Coriolis forces cancel each other out and the air moves in circles, forming a depression. Similarly, rather than travelling directly away from a region of high pressure, winds circle around to form an anticyclone.

And as these winds circle around, they also encounter another force. The air in a depression is prevented from moving inward, not only by the Coriolis force, but by its own momentum; air swirling around the centre creates a centrifugal force. The consequence is that when the air pressure is low it tends not only to be wetter but also windier, the reason why your umbrella is so likely to be blown inside out. In contrast, in an anticyclone, the momentum of the air actually helps it move away from the high-pressure region. The consequence is that not only is the weather drier when the pressure is high, but also the winds are lighter—which is why on hot sunny days Victorian ladies could get away with holding flimsy parasols.

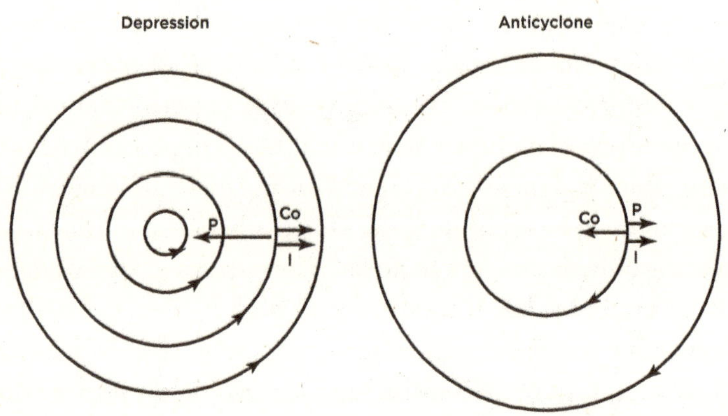

Pressure isobars and forces on the air in a depression and an anticyclone. In a depression, strong inward pressure, *P*, is resisted by a combination of Coriolis forces, *Co*, and the inertia of the air, *I*, so wind speeds can be high. In an anticyclone, the Coriolis force, *Co*, acts in the opposite direction to both the pressure gradient, *P*, and the inertia, *I*, of the air, so wind speeds tend to be much lower.

The lack of Coriolis forces at the equator result in very different weather from temperate areas and a total lack of weather systems. The sun is of course extremely strong, explaining why it evaporates so much water and why it rains so much. However, since Coriolis forces have no horizontal component in this region, the air merely moves upward, forms clouds, and produces afternoon downpours. I remember on my first trip to the rainforest of Sabah, North Borneo, being baffled as well as suffocated by the stillness of the air in the country that the locals call the "Land Beneath the Wind." It was a sharp contrast to the blustery storms and drizzle I knew so well back home in Manchester.

Things become different away from the equator, however; at latitudes above 4 degrees, Coriolis forces do become large enough for circulating storms to develop. The sun heats the surface of the sea and evaporates water from it, producing warmer, wetter air, which rises. Air moves in from all sides to replace the rising air at the centre of the storm, but is deflected by Coriolis forces into the characteristic doughnut of clouds that spins around the eye of a storm. You would not expect such a storm to grow to any size or strength because before it did so it would be carried back to the equator by the trade winds. However, near land the trade winds can be deflected towards the poles. The Caribbean and South China coasts deflect the trade winds northward, for instance, carrying developing hurricanes and typhoons with them. And as they travel, the sun continues to input energy into the system, and the winds move more and more rapidly in towards the centre. By this stage, inertial forces begin to far outweigh the effects of Coriolis forces in keeping the winds travelling in a circle. As the sun continues to input energy into it, the storm just gets stronger and stronger. The winds swirl faster and faster around its eye and form the hurricanes of the Caribbean and the Gulf of Mexico, the typhoons of Southeast Asia, and the cyclones of India. When they reach land these storms can devastate all before them, before losing power as they travel over the cooler, drier ground. And under the influence of Coriolis forces, they then veer even farther towards the poles, finally turning westward by the time they reach temperate latitudes, becoming deep depressions.

So the spin of the earth has created a planet on which most of the continents are bathed in air that is warm enough and watered by precipitation that is great enough to enable life to thrive on the bare ground of what would otherwise be just dry, rocky, lifeless land. It has created a wide range of habitats in which a bewildering variety of plants and animals can grow and reproduce. And between five and two million years ago in tropical East Africa, it produced a savanna habitat that was in the rain shadow of the growing mountain ranges of the

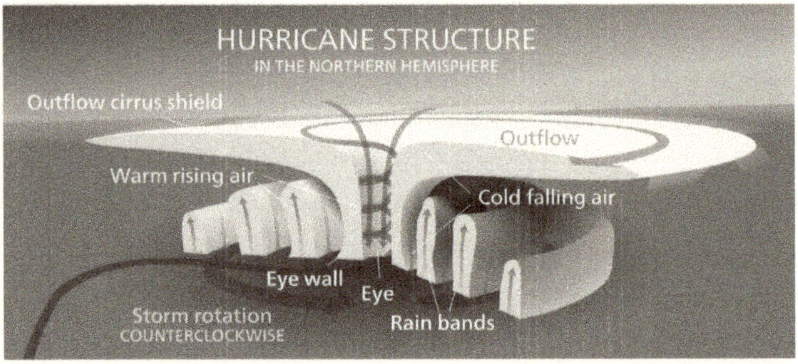

Structure of a hurricane. Evaporation of water in the warm sea below the eye causes air to rise, resulting in more air rushing in from the sides and being diverted into circular flow by Coriolis forces and its own inertia. As it reaches the top of the hurricane the air eventually flows outward and starts rotating in the opposite direction and releasing its water vapour.

Great Rift Valley, one in which an arboreal ape was forced to descend from the trees and become a terrestrial biped. Eventually that biped was able to travel from its homeland and reach almost every part of the world, colonising regions such as the Mediterranean and South Sea Islands that could almost be described as being like the Garden of Eden. It's true that we are beset with volcanoes, earthquakes, and hurricanes, but for the most part conditions on earth suit us just fine. Spin may not have given us the best possible of all worlds, but it has certainly given us one that is more wonderful than we could possibly imagine. It's a world that today, as we load the atmosphere with greenhouse gases, carbon dioxide, and methane, we are starting to change. We must act fast to reduce these emissions and store greenhouse gases so that we do not cause irreversible changes to the earth's climate and inflict massive ecological damage to our planet.

PART II

SPIN IN OUR TECHNOLOGY

CHAPTER 6

Spinning and Drilling

Our ancestors would not have known that we owe our existence on this planet to the way that it spins through space and rotates around the sun, or indeed that the earth is moving at all. Nor would they have had any formal knowledge about spin, or any aspect of mechanics, come to that. Nevertheless, as we shall see in this part of the book, people have been using rotating devices for thousands of years. They have designed and built a vast range of ingenious tools and machines, all of which are based on the principles we can see operating in the motion of the planets; most of them rotate about a single fixed axis. And our ancestors, seemingly intuitively, learned how to exploit the fascinating properties of spinning objects to construct our industrial world and raise our standard of living. And it all started in a very small way: in the home.

Among the earliest evidence of people using spin are the finds of whorls: small perforated disks made of stone, bone, clay, or metal that have been found in archaeological digs from the Neolithic period onward. They are common finds, but of all the grave goods unearthed by archaeologists, they must rank among the least celebrated. It is easy to see why. Small enough to fit neatly into the human grasp, they are far less spectacular than the axe heads, swords, and shields that the general public crave; and they are found not in the graves of great kings or warriors, but of women. Yet the whorls mark the birth of an important new technology—the rapid production of thread—and indicate that Neolithic women had a good grasp of rotational motion. It is their industry, after all, that gave its name to the motion that is the subject of this book—spinning.

From early in human history, people must have felt the need for long stretches of cord that they could use to tie things together: to haft stone blades to their spears and axes; to sew animal hides together to clothe themselves and roof their huts; and to fasten together the structural timbers of their houses. They must have used the stems of climbing plants and young tree saplings, and strands of leather, sinew, and guts from the animals they hunted. But solid lengths of cord have two disadvantages. First, they are all too easy to break. A small notch in a leather thong quickly turns into a crack that runs right through it, which is why the straps of your sandals can snap after very little use. Fortunately, this is not such a problem for many plant stems and mammalian muscle tendons, because the material of which they are made is split up into large numbers of isolated fibres. But even plant stems and tendons suffer from a second problem. The thicker a cord, the more rigid it becomes, so the harder it is to bend: to wrap it around small items, or tie it into knots. Neolithic woodworkers realised that they could bend coppice poles much more easily if they first twisted them so that the fibres sheared past each other and separated. This allowed each separate fibre to bend on its own, so the woodworkers could use twisted coppice poles rather like ropes.

But women also realised that they could make strong, flexible cords by joining lots of fibres together side by side and wrapping them around each other. One method of making a permanent cord in this way is to braid fibres together. Women have long used this technique to tie their hair into plaits and to make bracelets; such braided cords combine strength with flexibility. However, braiding is a time-consuming process and requires fibres that are all of equal length. Rather than braiding fibres together, it is quicker simply to twist them together into a multiple helix. Cord produced using this technique is not only extremely flexible, just like braids, since the fibres readily shear past each other when the cord is bent, but it has a second major advantage. When you pull on a twisted cord it draws the fibres together, and the friction between the fibres prevents them from sliding apart, so each fibre does not need to span the whole length of the cord. Since the failure of a single fibre has little effect on the strength of the whole cord, the technique also produces thread that is much longer-lasting. The overriding advantage is that this twisting technique allowed people to join even short fibres together, opening up the possibility of exploiting a range of materials, from the cellulose fibres in the stems of flax and the flowers of cotton to the keratin fibres in the wool of sheep, goats, and alpacas. It enabled people to manufacture strong, flexible threads.

Of course, thread is more perishable than stone, but keratin and cellulose fibres can survive reasonably well in dry environments such as caves or in the waterlogged conditions of a peat bog, so evidence of twisted thread dates back well into the upper Paleolithic period. The earliest twisted yarn found to date comes from the Dzudzuana Cave in the Republic of Georgia, where twisted flax fibres thirty thousand years old have been discovered. These threads were probably produced by rolling lengths of fibres together on the thigh, but this is a slow process and produces very irregular thread. As people started to settle down and use looms to make woven fabric they would have needed to speed up yarn production. Their solution was the invention of the spindle, a device that exploited the first law of rotational motion: that objects once set spinning keep rotating. The first spindles were probably just simple wooden sticks 8 to 12 inches (20 to 30 centimetres) long with a cleft at the top into which the end of the yarn was jammed. The spinner drew out a few inches of fibres from a bundle of material with the left hand, and then with the right rotated the dangling spindle between her finger and thumb, spinning it to twist the threads together. She would then stop the spindle and wind the new thread onto it like a cotton reel, and repeat the process, almost endlessly. In the new farming and pastoralist societies of the Neolithic period, it was women, especially young unmarried girls, who were lumbered with this repetitive and time-consuming task—hence the name "spinster" for an unmarried woman.

These young women were quick to make improvements to their equipment. They attached the ball of raw fibres onto a tall oar-like tool—the distaff—which they tucked under their left arms. This freed both of their hands to draw out the fibres and rotate the spindle. And they showed an intuitive understanding of rotational mechanics to keep their spindles rotating for longer. One way was to attach a wooden crosspiece to the bottom of the spindle. As well as preventing the thread falling off the bottom of the spindle, this made it hang more stably, and it also increased its moment of inertia, enabling it to spin for longer once set rotating. An even better solution was to attach a disk-shaped piece of stone—a whorl. Not only was the stone heavier than a wooden crosspiece, but more of its weight was concentrated farther away from the axis of the spindle, giving it a higher moment of inertia, so it proved even more effective at keeping the spindle spinning. This development allowed women to produce finer, more consistent thread much more quickly. And since they had to spend so much time on this repetitive task, not only in this world but—as the finds of spindle whorls in their graves suggests—presumably the next, women ensured that the whorls they used were good to look at as well as to use. Later whorls that were made from clay and, in Roman times, lead were beautifully decorated with a wide range of geometrical patterns and depictions of flowers. They thus went one step further than even the Victorian designer William Morris advocated, and made sure that their most prized possessions were both useful and beautiful.

Since their spindles dangled down from the end of the thread and were weighted towards the bottom, they hung down vertically and were stable, even when they were not spinning. However, the women, or maybe their children, must also have noticed a side effect of the rotation of their spindles—its ability to stabilise them even when they were not hanging from a thread. If a spindle was set in motion and dropped onto a smooth floor, it would not only keep on spinning but also remain upright, apparently resisting the inexorable grip of gravity. They had independently discovered that spinning objects are stable. The result was the invention of the first spinning tops, objects that have been favourite toys ever since, fascinating generation upon generation of children and adults alike.

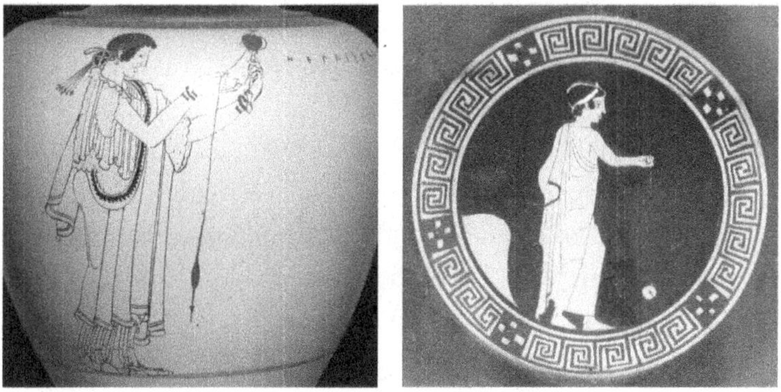

Depictions on Greek pottery from the fifth century BC of a woman spinning thread (*left*) and a youth playing with a yo-yo (*right*).

The first spinning tops were probably twirlers, barely modified from spindles and composed of just a shaft and a whorl. They would have been set spinning by rolling the shaft between the finger and thumb, and improved merely by sharpening the bottom of the shaft into a point to reduce friction. But later a whole host of tops were designed to delight children's minds and sharpen their motor skills. There were throwing tops, launched by wrapping thread around their portly bodies and then throwing them to the floor; and whipping tops, kept spinning by hitting them using a leather cord on a stick. Tops first appear in the archaeological record in ancient Mesopotamia, where a clay top has been found that dates to around 3500 BC, and they were common in ancient Egypt; Tutankhamen's tomb contained a beautiful wooden top with inlaid ivory and ebony decoration. By classical times there were several types of rotating toys. As well as twirlers, throwing tops, and whipping tops, there were also whizzers, disks resembling large buttons that are spun using two cords that run through two holes near the centre of the disk and that are wrapped around the hands.

The player moves the hands in and out, which alternately winds up and unwinds the two cords, thus spinning the disk. And there were even yo-yos: sets of paired disks joined by a short axle that are set spinning by a cord that is wrapped around the axle. We know this not only because of archaeological finds, but because these spinning toys are also depicted on the graceful red figure pottery that is one of the artistic marvels of classical Greek civilization. A surprisingly high proportion of the vases that were produced during Athens's golden age during the late fifth century BC show women playing with tops, whizzers, and yo-yos. They outnumber those depicting the women carrying out useful tasks, like the Fate Clotho spinning thread or Odysseus's long-suffering wife, Penelope, weaving cloth. The emphasis on play is unlikely to be an indication of the actual lifestyle of Athenian women, however; rather it is probably yet another example of men's eternal, if irrational, belief that when their husbands are out of the house, married women lead a life of leisure.

The stone whorls of the Neolithic period also point to another way in which these people made use of rotation: to make drills. Like the heads of hammer axes and items of jewellery such as rings and beads, many of the whorls were perforated with perfectly circular holes—some of the first examples of precision engineering. Experimental archaeologists have shown that hard materials such as shell and stone can readily be perforated using simple drills made from straight sticks that can either be tipped with stone points or coated with abrasive sand. If the stick is rotated quickly by rubbing it between two hands while simultaneously applying a downward pressure, the point gradually wears the material away, leaving a circular hole. Artifacts in stone, antler, bone, and shell that have been pierced in this way to make jewellery have been found around the world from as long ago as the upper Paleolithic period. And people quickly developed ways of improving the performance of simple hand drills.

The bow drill appeared as long as ten thousand years ago. Hunters found that they could wrap the string of a modified hunting bow around a wooden rod and move it back and forth, like a violin bow, to rotate the rod at higher speeds than they could by rubbing it between their hands. And they could drill the hole faster by pressing down on the rod with a stone weight. Alternatively, the hunter could attach a stone whorl to the rod, and by adjusting the tension in the bow, ensure that the drill was only accelerated on one stroke, exploiting the angular momentum of the whorl to keep the rod spinning in the same direction, just like in a modern drill, rather than back and forth. Bow drills spread quickly around the world and were used for a variety of tasks. A tomb in Mehrgarh, Pakistan, dated between 7000 and 5500 BC, contains teeth that have been drilled with small holes—the first dentistry—while bow drills are frequently depicted in the tombs of ancient Egyptians, where they are being used by carpenters and bead makers.

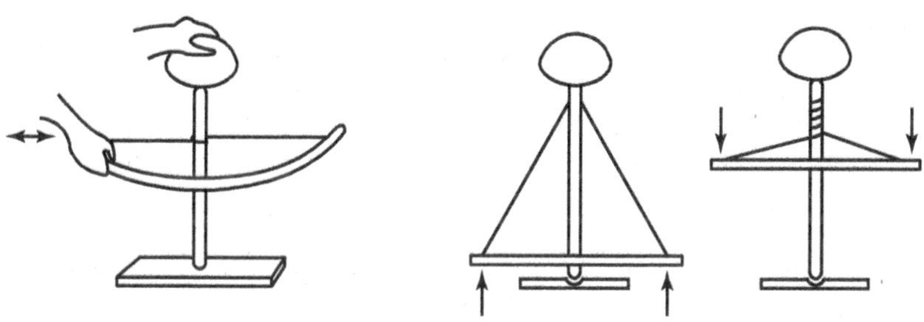

Two early types of drills. In the bow drill (*left*), the drill bit is spun by wrapping the string around the drill and sweeping the bow back and forth, like on a violin. In the pump drill (*right*), the drill bit is inserted through a hole in the board and is spun by moving the board up and down.

Two further designs improved on the efficiency of the bow drill: the cord drill and the pump drill. In the cord drill the bow was replaced by a cord that was attached to a groove at the top of the drill, just like the thread of a spindle. The two ends of the cord could then be wound around the rod and pulled outward and downward to simultaneously set it spinning and push down on the drill bit. The technique was perfected in the pump drill, in which the two ends of the cord were attached to the ends of a narrow board that had a hole drilled through its centre, through which the rod was inserted. To rotate the rod, the user moved the board up and down, alternately wrapping and unwrapping the cord, just as in the cord drill. The advantage of this arrangement was the greater control the board gave the user in orienting the drill; the rod was constrained both at the board and lower down within the hole that it was drilling, rather like the axle of a wheel, so the hole could be drilled more precisely. Cord and pump drills are still used around the world, particularly by hunter-gatherers, who appreciate their lightweight design and the fact that they can be constructed using a couple of small pieces of wood in just a few minutes. This makes them the automatic choice, not just to drill holes, but to also light fires. All the fire lighter has to do is place a pile of wood shavings into the well he or she has drilled into the lower board. Friction between the stick and the board heats the well up to the temperature at which the wood shavings spontaneously combust.

Of course, the wooden parts of drills quickly perish, so apart from their whorls, their remains are rarely found in the archaeological record. Consequently, we do not know how or when the first fire drills were made. However, recent Neolithic finds from Israel show that the fire-drilling technique had been perfected and commercialised as long as eight thousand years ago. What the archaeologists actually found were cylinders of clay 1.2 to 2.4 inches (30 to 60 millimetres) long and 0.5 to 0.6 inches (12 to 15 millimetres) in diameter, coming to a blunt apex at one end. The circular grooves incised around the apex suggested that they had been spun around in wooden chambers, as in the fire drills of modern hunter-gatherers, while oblique fractures of the other end suggested that they had been loaded and had failed in torsion. The cylinders must have been attached to the tip of a bow or pump drill, and had been produced as part of an organised industry: the first match makers.

Close examination of early grave goods, therefore, shows that both women and men exploited spin to make a major contribution to the Neolithic standard of living. Sitting by the fire they had lit, wearing the necklaces they had made with their drills, spinning the yarn they needed to make clothes with their spindles, and letting their children play at their knees with their spinning tops, spin would have made everyone's life warmer, cozier, and more fun. And as we shall see, using these simple spinning techniques was just the first stage in the development of the technologies that we have used ever since.

CHAPTER 7

The Underwhelming Wheel

One of the facts of life that have made it so hard for people to believe in Newton's first law of motion is that whenever you stop pushing something along the ground—a chair or table, for instance—it does not carry on moving as he claimed, but stops. Friction rapidly decelerates it and inertia seems to hold little sway. But for spheres and cylinders this is not the case. Early people would have noticed that many fruits and nuts, and the whorls of their spindles, would have carried on rolling along the ground long after they had released them. They would also have noticed that the rolling whorls, like modern coins, would have stayed upright long after they would have fallen over if they had not been set in motion. Nowadays it is quite easy to explain both these phenomena. A rolling disk has two motions: the translational motion of its centre and a rotational motion about its centre, which means that the bottom of the disk moves backwards at exactly the same speed as the centre is moving forward. Consequently, the rim does not slide over the surface and the forward motion of the disk is unimpeded by friction. The rolling disk is also stabilised due to the phenomenon of precession that we examined in chapter 3. If the disk starts to lean to one side, gravity will tend to turn it farther to that side, but as the disk starts to fall over, the spin will cause the disk to precess, rotating it about its vertical axis, so that it turns into the lean. The centrifugal force on the turning disk will then oppose the overturning moment and keep the disk from falling farther. It will carry on rolling in ever decreasing circles until it finally falls over.

But even though early people were unable to explain the mechanics, they would have realised that rolling circular objects along would be easier than sliding them along the ground. It was this that must have stimulated them to invent what is conventionally regarded as "man's" most important invention: the wheel. Indeed the emergence of the wheel as a method of improving transport is nowadays seen as the key landmark in the rise of civilization. So it is important to ask how and when the wheel was developed, how wheels work, and to investigate just how useful early wheels actually were.

Most children's (and indeed adults') books will tell you, without giving anything in the way of evidence, that the precursors to the wheel were simple logrollers that the ancients used to transport huge stones. The idea is that the stone was laid on top of a series of logs that were set parallel to one another on the ground. The stone was then pulled along with ropes, while the rollers moved forward at precisely half the speed of the stone, and reduced its resistance to motion. This theory has over and again inspired experimental archaeologists to attempt reconstructions of the supposed feats of ancient architects, endeavors that have proved irresistible to TV documentary makers. It's not hard to see why; these experiments focus on the most spectacular feats of the ancients: how the builders of Stonehenge moved the four-ton bluestones a hundred and fifty miles from Pembrokeshire to Wiltshire; how the ancient Egyptians transported huge limestone blocks from quarries in upper Egypt to the site of the pyramids at Giza; and how the pre-Inca Tiwanaku civilization moved stones from Lake Titicaca to the site of their capital, a few miles south.

Unfortunately, though simple in principle, moving stones using wooden rollers is actually extremely impractical and dangerous. For a start, the engineers would have had to cut up large numbers of perfectly circular logs of identical diameters. It would be hard to find enough suitable trees for the purpose, even in a well-wooded landscape, let alone the deserts of Egypt. And stone axes are so poor at cutting across the grain that it would be hugely time-consuming to cut the trees into logs; even if they had done so the logs would be pencil-shaped at both ends. But it is when it came to the rolling process that their problems would really have started. Even today, when tree trunks can be readily cut into logs with chain saws, reconstructions of stone moving using rollers are hardly unqualified successes, and for good reasons. The teams pulling the sleds, often reinforced by burly rugby players or tug-of-war squads, are certainly able to get them moving. But there has to be a second team of volunteers to pick up the logs that roll out behind the stone and replace them in front of it. Hauling large logs around in this way is highly strenuous and incredibly dangerous. It's virtually impossible to perform the task accurately and quickly enough, especially if the stone is moving along continuously. The process inevitably becomes a very stop-start affair, greatly reducing both its speed and efficiency. Other downsides are that the stone is all too apt to slip off the side of the rollers, while the rollers themselves may jut up against projections such as stones or dig into soft ground. The final problem is that since the rear of one roller moves in the opposite direction to the front of the one behind it, if they touch, the two logs will jam up against each other, bringing proceedings to a juddering halt. Consequently, after a few days of trials, the archaeologists usually give up; they regard their operations as a qualified success that "proves the principle," but they hardly ever move stones more than a few hundred yards. These experiments certainly don't inspire confidence that people could transport huge objects hundreds of miles across country using rollers.

In fact, the ancients probably floated their stones most of the way to their destination—along the Nile, across Lake Titicaca, or up the River Avon—using the rafts and plank ships that we know prehistoric people were adept at building. And they probably moved them the final few hundred yards across country by mounting them on sleds and sliding them along. This technique is frequently depicted on the walls of Egyptian tombs, where it is shown being used to transport vast statues of the pharaohs. Many recent experiments have demonstrated that the drag of a wooden sled can be sufficiently reduced using lubricants to allow a reasonable team to haul it along, without the chaos of using rollers. The Egyptians poured water onto the sand in front of the runners of their sleds, while the builders of Stonehenge probably lubricated their sleds with animal fat, and the architects of Orkney's stone circles dragged the stones across slippery kelp.

Bearing in mind how impractical rollers are, it's unlikely that they were used at all, so they could not have been the inspiration for the invention of the wheel. Instead, it seems much more probable that wheels were developed by scaling up the smaller devices—spindles and drills—that were already being used in the home. As we saw in the last chapter, the principle of the axle—a slender rod that rotates while being held within two housings—was already familiar to Neolithic people, in the form of the pump drill. In these tools the spindle rotates freely and stably not only within a hole in the handle but also within the hole that the tip has drilled. Moreover, bow drills were often already furnished with a wheel-like disk, the whorl, to keep them moving. Creating an apparatus with two wheels on a constrained axle would simply involve adding a second whorl and turning the apparatus on its side. The way in which this arrangement enables an object to roll along the floor would have been apparent to any woman with a spindle or any child with a twirler.

Of course, a wheel does not roll with the complete freedom of a ball or disk, but because the wheel's rotation is constrained only by having its axle harnessed within two housings, the apparatus would have been able to travel relatively easily across the smooth beaten earthen floor of an early hut. It would be easy to lubricate the axle, and since the friction on the axle acts so close to its centre of rotation, the force resisting the movement at the rim of the wheel will be far smaller. It's not surprising, therefore, that children's toys that were furnished with wheels have been found not only in the Old World, where full-size wheels emerged, but also in the New World, where they did not. The Incas, Mayans, and Aztecs all made a range of toy animals that ran on four clay wheels: jaguars, monkeys, dogs, and even alligators!

The fact that in the New World the principle was never developed and scaled-up to produce a useful wheeled vehicle was probably due to the difficulty that people using Stone Age technology would have had in constructing practical wheels from wood. You might think it would be easy to make a wooden disk; all you would need to do is take slices from the trunk of a tree. However, thick stone axe blades are simply unable to cut wood across the grain in the way that is required, unlike modern steel band saws. And even if it had been possible to cut out a disk of wood from the trunk of a tree, it would not have made a practical wheel. Wood is weak across the grain, so the wheel would have broken easily. And because wood shrinks more in the tangential direction as it dries out than radially, the disk would split and a pie-slice-shaped groove would open up. In the Old World the first wheels were not made until the advent of the Bronze Age in the fourth millennium BC, when the new sharp metal tools, axes, adzes, and chisels would allow woodworkers to cut timber precisely across the grain and hence make the first truly accurate joints.

Wheels first appeared in Eastern Europe and the Near East, and the first wheelwrights constructed their wheels by joining together two, or most often three, planks of wood. They linked the planks together side by side using tongue-and-groove joints, and they strengthened the joints by cutting grooves across the wheel's surface and inserting wooden battens into them. The wheels they produced in this way would have been small, heavy, and quite weak, but they would at least have been usable. The first wheeled vehicles, two-wheeled handcarts, were produced in the middle of the fourth millennium BC, and four-wheeled wagons soon followed. Farmers also modified their carts so that they could be pulled by oxen, by harnessing them to the same sorts of yokes and shafts that they were already using for their primitive scratch ploughs. Carts and wagons gradually became more common and were soon being used not only to transport goods around farms and to and from local markets but also to make deliveries in the towns and cities of Mesopotamia. Wheelbarrows, furnished with just a single, central wheel, were used in China from the second century BC not only in agriculture but also as a form of high-class transport, with people sitting either side of the wheel. Wheels, it seems, were about to revolutionise transport in the Old World.

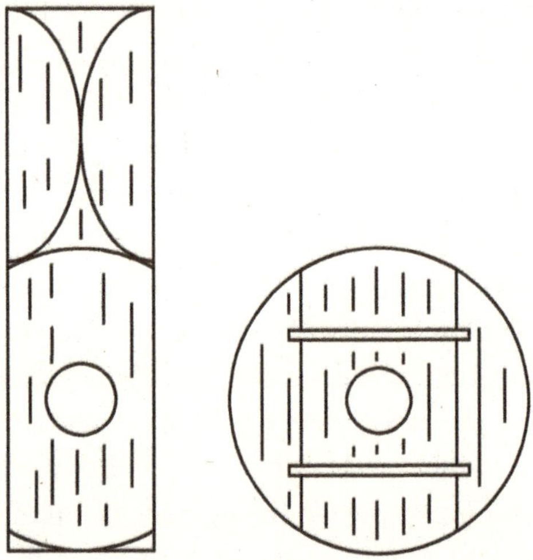

Construction technique to make a Bronze Age wheel. The three pieces were cut from a single plank of wood, joined side to side using tongue-and-groove joints, and fixed together using dowelling crosspieces.

Indeed, there is no doubt that wheeled vehicles have some advantages as a means of transport. For a start, the wheels hold the body of the vehicle above the ground, so unlike human rucksacks or the paniers laid across pack animals, in wheeled vehicles no energy is needed to support the goods, which in itself can be backbreaking and energy-consuming. And they should theoretically suffer from far less friction than sleds. But it is all too easy to overstate the importance and the advantages of wheeled vehicles. They need a hard, flat surface in order to roll easily. Wheels quickly get bogged down in soft mud or deep snow, unlike sleds, which transmit their load to the ground over a much wider area. And on rough, stony ground wheels subject vehicles to a constant jolting as the rim encounters projections. This slows them down since it uses up energy to repeatedly decelerate and accelerate and raise and lower the vehicle. This explains the popularity of sleds and sleighs in Russia, and of travois, simple sleds made from two jointed poles, on the Great Plains of North America. One way of reducing both of these problems is to make the wheels with a greater diameter, a factor that must have helped drive the development of the spoked wheel, which is far lighter and stronger than the primitive plank wheel and can be made much larger. Right up to the twentieth century, farm carts and hay wains were being built with huge wheels over 8 feet (2.5 metres) in diameter. Even so, wheeled vehicles could only be moved easily on smooth, firm, and level ground. In Scotland, wheeled vehicles were rarities, even in the Lowlands. Goods were transported locally on sleds; and until the latter part of the eighteenth century there was no wheeled transport at all between Scottish towns, even between Glasgow and the capital, Edinburgh.

Another way of improving wheeled transport, of course, was to construct roads with a smooth, hard surface. However, roadbuilding is difficult across boggy mountainous areas, and near impossible in sand, so throughout the ages merchants have preferred to carry their goods across hilly country using horses, donkeys, and mules, while camels were preferred to carry goods across desert sands such as the Sahara. And even the most famous long-distance trade route, the Silk Road, was not actually a road as we would imagine it, but a series of trails navigated by caravans of pack animals. The New World civilizations thrived for hundreds of years without using wheels at all, and built cities and temples as awe-inspiring as any in the Old World. They did so by relying on human porters and on pack animals such as llamas to carry their trade goods. Indeed, using wagons off-road was seldom successful. When American settlers migrated across the continent in the nineteenth century along the Oregon Trail, their wagon trains repeatedly foundered as they attempted to cross the Rocky Mountains. And the Great Trek of the Boers across South Africa later in the century was greatly slowed by the rugged ground.

Even in temperate Europe, long-distance roads were rare. In ancient times, it was only during the rule of the highly organised Roman Empire that a comprehensive road network was built and maintained, and this was used not so much for trade, but to allow the army to march quickly and easily to trouble spots. After the demise of the empire, the road system of Europe quickly fell into disrepair.

After all, it was far easier and cheaper to transport goods by water than across country, which was a major reason why large cities were almost always located on the coast or on the banks of a navigable river. It was not until the end of the eighteenth century that the first industrial country, Britain, developed roads that were not dangerously rutted in summer and waterlogged mud baths in winter. Civil engineers such as John McAdam and Thomas Telford pioneered cambered roads that were constructed of carefully sorted stones. But such engineering works are expensive, so the tolls on the new turnpike roads were too high for anything but the post chaises and carriages of the wealthy. Cattle were still marched hundreds of miles from as far afield as the Highlands of Scotland along drovers' roads to the meat markets of London and other cities. And until the middle of the nineteenth century, consumer goods were often transported directly to customers on packhorses; packmen travelled long distances on narrow pack roads, which were basically just single lines of flagstones. The packman, Bob Jakin, is an important character in George Eliot's *The Mill on the Floss*, a novel set in rural England as recently as the 1830s. Coach travel was slow and dangerous, even along the developing network of turnpike roads. The novels of Jane Austen consistently rail against the slowness, cost, and danger involved. In *Mansfield Park*, for instance, it takes a whole day to travel from Portsmouth to Oxford, nowadays a mere ninety-minute trip along modern divided motorways, leaving everyone bruised and tired. In *Persuasion* it takes three and a half hours to travel the seventeen miles to Lyme Regis, and *Sanditon* starts with the protagonists being overturned in a carriage accident.

Over time, carriage builders endeavored to overcome two of the disadvantages of coaches: their slow, bumpy ride and their difficulty in turning corners. The former problem was mitigated by adding some suspension that shielded the body of the coach from the rise and fall of the wheels over bumpy ground. At first, this mostly involved coach builders hanging the body of the carriage from long leather straps or chains, but the engineering started to improve in the seventeenth century when members of the Royal Society started experimenting with a range of springs. But it was not until the beginning of the nineteenth century that Obadiah Elliott invented the elliptical leaf spring; the axle was mounted onto the centre of a curved iron plate that was attached to a series of ever shorter plates, while the two ends were attached to the body of the carriage. The ability of the spring to bend provided suspension, while friction between the iron plates provided some shock absorption, damping the oscillations down. The only downside was that the damping was hard to control, so the recoil of the springs was all too apt to toss passengers into the air. The lightest and most manoeuvrable carriages, though rather unstable and uncomfortable to ride in, were two-wheeled gigs and curricles, pulled by one or two horses respectively. The equivalent of modern sports cars, they were ideal vehicles for young men to impress fair maidens, as Jane Austen recounts in *Northanger Abbey*.

Getting over the second problem, making carriages more manoeuvrable, proved rather harder. It is easy for a two-wheeled cart to turn a corner; the wheels just have to rotate independently on the axle so that when the cart reaches a corner, they can spin at different speeds. However, for four-wheeled wagons and coaches, even the luxury barouche-landau owned by the Suckling family in *Emma*, turning a corner was a much greater challenge. In a four-wheeled wagon with fixed axles, even if each wheel is free to spin independently, one set of wheels has to be dragged across the ground to enable the wagon to turn, which can prove almost impossible. One method of overcoming this problem, first devised in Hungary in the fifteenth century, was to mount the front axle of the carriage on a pivot that was linked to the horse shafts so that the two front wheels could turn around the corner. This certainly helped, so most carriages from the sixteenth century onward were furnished with "Hungarian steering." But this design had disadvantages. Foremost, it meant that the front wheels had to be smaller than the back ones to enable them to swing under the chassis of the carriage, which made the ride bumpier. The second problem was that as the axle turned, the front wheels moved inward, reducing the width of the wheelbase and making the carriage less stable. Carriages were consequently prone to overturn at corners.

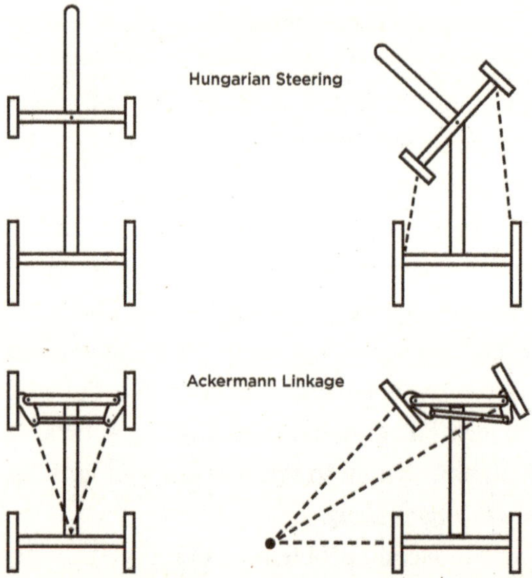

A comparison of Hungarian steering and the Ackermann linkage invented by Erasmus Darwin. In Hungarian steering the front wheels have to be small and turning reduces the width of the wheel platform. In Ackermann linkage, the inner front wheel turns more than the outer one, so that both wheels are at right angles to the centre of rotation and the wide platform is maintained.

Together, with the expense of keeping two or even four horses, these problems meant that four-wheeled carriages were used only on the best roads: along turnpikes, and in cities. Even in town many people, such as the ladies of Elizabeth Gaskell's *Cranford*, preferred to be carried by footmen in sedan chairs well into the nineteenth century; these gave a far smoother ride and were much cheaper. Meanwhile, the people who needed to get about the countryside on business—farmers, lawyers, and doctors; people such as Tertius Lydgate and Caleb Garth in George Eliot's *Middlemarch*—travelled between appointments on horseback, whatever the weather. It is notable that one doctor who was unable to ride—being too fat—was the first person to design a satisfactory steering mechanism for carriages. Erasmus Darwin, the grandfather of Charles Darwin, the pioneer of evolution, was not only a successful doctor but also a polymath. He was the composer of best-selling poems about the love lives of plants; a pioneer in both the antislavery and feminist movements; and the designer of a speaking apparatus. To overcome the dangers and lack of comfort of his carriage, he designed a steering system, which was later patented in 1818 and known as the Ackermann linkage. In this mechanism, the steering pivots angle inward and are joined by a tie-rod that can be moved to the left or right by a rack-and-pinion system. Steering the wheels turns the inner wheel by a greater angle than the outer one, so that each wheel follows parallel to its line of travel, and neither wheel moves inward, so the stability of the carriage is unimpaired.

But though wheels were of limited use in civilian life, history books often celebrate their role in warfare. By 2000 BC, the Sumerians were already incorporating solid wheels into a novel weapon of shock and awe: the battle chariot. The very name seems to invoke terror, and official bas-reliefs always imply that the battle chariot was a supremely effective weapon, capable of transporting an elite warrior to any point of the battlefield. The truth was probably rather less impressive. The original Sumerian battle chariot, for instance, was a small wooden and wickerwork cart, fitted with two small plank wheels, pulled by asses, and holding just two people: a driver and a spear- or bow-wielding warrior. And though chariots have long fascinated small boys and archaeologists, they could hardly have been the terror weapon they have been portrayed. Back in the 1970s I remember watching a BBC *Chronicle* documentary that followed a team of experimental archaeologists who were attempting to reconstruct Sumerian battle chariots. The tone of the documentary was earnest, but the films of these vehicles being pulled along by donkeys were more comic than terrifying; the chariots were cumbersome, slow, and hard to manoeuvre, bumping ponderously along as the donkeys trotted in front of them, even on the smoothest of ground.

The later chariots of the ancient Egyptians and the Homeric Greek heroes would have been more impressive, since they had stronger, lighter spoked wheels, which provided some suspension. They were also pulled by more formidable ponies that could be directed using the newly invented bridle and bit. But even these vehicles would have been more for show than real practical utility. They would certainly have enabled wealthy aristocrats to be taken to the heart of battles without getting out of breath, but they would only have been practical on smooth, level battlefields, and so only suitable for stylised set combats. By classical times, chariots were obsolete, since larger horses had by then been bred that acted as the mounts of armed soldiers. Cavalry squadrons were far more manoeuvrable and effective fighting units, as was demonstrated by Alexander the Great in his victory over the Persians at the Battle of Gaugamela in 331 BC. The Persians spent the night before the battle smoothing the battleground to make it suitable for chariot warfare. However, when the Persian chariots charged Alexander's infantry, the phalanxes simply opened up, allowing the chariots through, so that the lightly armed Macedonian hypaspist troops could pick the charioteers off from behind. Meanwhile, Alexander's crack cavalry ran amok among the remaining Persian ranks. The famous mosaic of Alexander, mounted on his steed Bucephalus, bearing down on the chariot of the alarmed Persian emperor Darius illustrates the demise of such fighting vehicles. The same fate was later meted out to the chariots of the Britons by the invading Roman forces in 55 AD. Despite the legendary courage of the Britons' leader, Caractacus, and the rotating scythes on the axles of the chariots that capture the imagination of bloodthirsty children to this day, the Roman legions simply let them through their ranks and dealt with the charioteers one by one.

Wheeled vehicles were not seen again on the battlefield for two thousand years. Their only involvement in the intervening centuries was to form the wagon train that armies used to transport equipment and camp followers, and later the two-wheeled horse-drawn gun carriages that they used to move cannons. Both sorts of vehicles were slow-moving, prone to getting bogged down, and highly vulnerable to attack. Wheeled vehicles took no part at all in the invasion of the New World by the Spanish conquistadores; their ships would have been too small to transport them across the Atlantic, and in any case there would have been no proper roads for them to travel along. As we shall see, wheeled vehicles were not to reemerge in warfare until the development of tanks in the First World War. The simple fact is that, until a hundred years ago, wheels were used even less in the military than they were in civilian life.

CHAPTER 8

Shaping Up

Though wheeled vehicles proved to be a disappointment, the principle on which wheels were designed—having a circular disk that rotates around a fixed axle—proved to be a technological game changer. As we shall see, it has since acted as the basis of almost all human machinery. Bronze Age people would have found that if they overturned their carts and spun their wheels around, these rotating disks have a motion that is paradoxical—at the same time fast-moving, but also stationary. It's a motion that even babies find fascinating, hence the popularity of spinning toys in cribs and strollers. The rotating wheel combines two usually mutually exclusive attributes: it possesses a large store of kinetic energy that can be used to perform work, and each part of the wheel has its own precise, predictable motion. The rim of the wheel, for instance, is always exactly the same distance from the centre of rotation and passes exactly the same points in space on each revolution. Consequently, just as early bow drills and pump drills used their axles to bore precisely circular holes in stones or teeth, so people realised that rotating wheels could be used to shape objects on a much larger scale, and process a much wider range of materials.

The first, and perhaps best-known rotating shaping device, was the potter's wheel. People had been forming and firing clay pots since the Upper Paleolithic period, and after the invention of farming, clay vessels had become much more common in the homes of the now sedentary farming communities. But it is a tricky and time-consuming business to shape floppy blocks of clay into thin-walled vessels. One common method early potters used was to roll the clay into long ropes and then build their pots upward by coiling the ropes together into a close-bound helix. The coils could then be smoothed together to produce a pot with a flat surface. This works, but it takes a great deal of time and skill, and the potter continually has to move the pot around to get to each side. Towards the end of the fifth millennium BC, Macedonian potters found that they could speed the process up by shaping their pots on a turntable that was mounted on a vertical wooden post. However, it was not until around 3000 BC, about five hundred years after the emergence of the first wheeled vehicles, that the first true potter's wheels appeared: small round tables that were mounted on a vertical axle. Potters could spin the table around rapidly, enabling them to use the energy in the spinning table to shape their clay into perfectly circular bowls, pots, and plates. And once the shape had been produced, they could decorate their pots by holding the blade of a knife against the side to incise circumferential grooves, or touch the pot with a brush laden with coloured glaze and spin it around, to produce circumferential stripes. The basic methods of making and decorating pottery had been perfected and have not been bettered ever since, even by industrialization.

The method of incising grooves in pottery probably acted as an inspiration for the development of a second machine that has proved even more important than the potter's wheel: the lathe. The advent of iron smelting around 1000 BC enabled blacksmiths to produce tools with sharp, resilient cutting edges, chisels and knives that could readily cut wood across the grain without wearing down too fast. To create finely shaped wooden objects with perfectly circular cross sections, craftspeople needed simply to clamp blocks of timber at their two ends onto metal bearings mounted onto a rigid frame so that they could be spun around rather like the axles of a cart. The lathe operator could then hold the point of a chisel against the rotating wood, gradually peeling material off to produce a whole range of circular objects—not just long ones such as tool handles, the axles and spokes of wheels and the legs of chairs and tables, but also complex vessels such as cups, bowls, and even plates. Wood turning quickly became a profession, with several specialties, from the bodgers who worked in woodlands to produce simple poles to town-based craftspeople, such as the cup makers, who worked in Viking York's Coppergate, or "Cup Street." Until the eighteenth century, when vast factories started to make cheap, colourful pottery, most people ate and drank off wooden plates, bowls, and cups, which were far more robust than china ones and could last them a lifetime.

A rotating wheel could itself be used as a shaping tool. People realised that if a wheel was made of a hard, abrasive substance such as stone, and a wooden or metal object was held up against it, the object would be gradually ground away. So a rotating stone wheel could be used to smooth wood, or to sharpen the blades of metal tools such as hoes and knives; sanding and grinding machines were born. Travelling knife grinders were common street traders right up until the middle of the twentieth century, roaming residential areas to hawk their services. And if the wheel was a slender metal disk, it could be provided with teeth around its edge and used as a cutting tool. As iron wheels became available they started to be used as circular knives that could cut through stone, or as circular saws that could cut through wood.

But there was an even more important material that the early farmers needed to process: grain. Neolithic farmers found that annual cereal grasses such as wheat, barley, and rice produce large harvests of seeds that are full of nutrition. One of the easiest ways of preparing such seeds was to boil them to produce a range of broths, porridges, and gruels. However, to enable the seeds to expand and absorb water, the cooks first had to break up the grains to allow them to escape from their seed capsule. One way to do this was to put the grains in a solid stone mortar and pound them with a heavy pestle, but this proved to be a very energy-intensive process. A better method was to place the grains on a flat stone and roll a heavy "rolling pin" back and forth over them. This rolling technique was perfect for producing flattened grains, such as rolled oats, that people could make into porridge, or pearl barley that they could boil in water to make into a broth; further rolling broke up dry grain to produce coarse fragments that could be mixed with milk or water and used to prepare puddings such as semolina.

Once the principle of the rolling pin had been invented, women quickly found other domestic uses for it, most notably in smoothing and drying cloth. The first mangles, which were commonly used from the Middle Ages onward, were simply wooden cylinders around which the clothworker could wind a piece of cloth. They would then roll the cylinder back and forth on a clean, flat surface using a wooden "mangle board" so that they did not need to touch the cloth. This squeezed out creases and removed water from within the weave. By the nineteenth century, mangles started to be improved by using paired rollers, which were held parallel on adjacent axles, leaving just a narrow gap through which they could squeeze the cloth. As we shall see, this method also proved to be useful for much heavier-duty work.

Rolling has limitations as a means of food preparation. The large particles it produces are best boiled, which makes for a monotonous diet. Boiled meals also contain so much water that they are hard to preserve and difficult to carry around. To make a longer-lasting and more portable meal, the grains needed to be ground up into a flour that could be mixed with water and fat to form a dough that could then be baked to produce a range of breads. Neolithic people quickly came up with the technology to do this, the saddle quern. They placed the grains on a shallow depression on a large stone, known as the quernstone, and crushed them by rubbing a heavy convex stone, known as a muller, rubber, or handstone, back and forth over the grain. Once the flour had been ground and mixed into a dough, Neolithic women could work it and shape it into a range of flatbreads using rolling pins, or throw it spinning into the air, before baking it.

Grinding grain between sliding stones comes at a cost, however. For a start, with a saddle quern, it was very hard to hold the muller at the same orientation, resulting in severe hand strain. Another big disadvantage is that using a quern was a backbreaking, laborious process. The women, to whom this unenviable task was delegated, had to spend hours every day kneeling over their querns, and rocking back and forth, using large amounts of energy not just to grind the grain but also to accelerate and decelerate the heavy handstone and the whole of their upper bodies. The skeletons of Neolithic women show not only the heavy wear on their skeletons that resulted but also the damage this did to the joints of their feet. Since they were kneeling forward, their toes had to be angled up at an unnatural angle, and so Neolithic women were frequently the victims of severe osteoarthritis.

The solution that reduced at least some of their workload was the rotary quern, a tool that, like the saddle quern, was made from two heavy stones. In this case, however, the upper stone was shaped like a ring doughnut and was rotated above the lower one using a wooden handle inserted near its rim. It was held in place by a projection in the centre of the lower stone that acted like an axle. Women could pour the grain into the quern through the central hole and rotate the upper stone, grinding the grain, which simultaneously worked its way outward as it was progressively crushed. The rotary quern was invented around 500 BC and quickly made its way across Europe, reaching Highland Scotland by around 200 BC, where it continued to be used until modern times. Since the motion of the upper stone was continuous, the rotary quern was a far more efficient device than the saddle quern, and acted as the inspiration for later water and windmills, as we shall see in the next chapter.

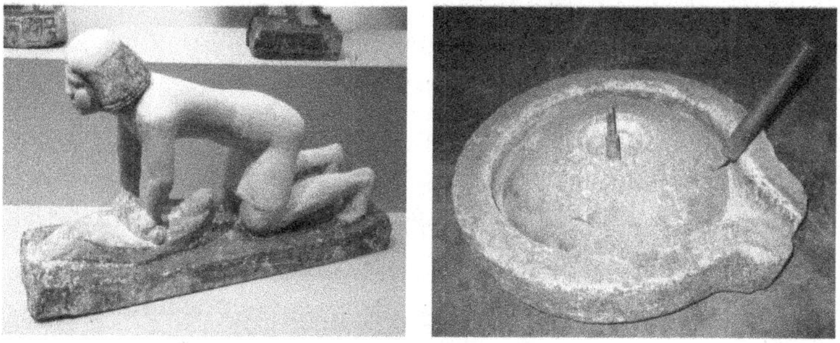

Machines to grind grain. The saddle quern (*left*) demands much greater muscular effort than the rotary quern (*right*), which merely needs its handle turned.

Even though rotating machines need far less energy to operate than reciprocating ones, since they do not need to be continuously accelerated back and forth, they still need to have an input of energy to overcome friction and power their operations. People soon came up with a range of different methods to drive their rotary devices, methods that we still use. The obvious way to do this was to push them around, an operation that can be made far easier by providing a handle near the edge of the wheel. Effectively, in this case, the operator uses their arm as a crank, providing the energy at the shoulder simply by swinging the upper arm back and forth, a reciprocating angular motion, and allowing their elbow to bend passively to allow the hand to sweep around with the handle in a circle. This works well for some operations, such as grinding grain, but it monopolises one arm, uses a lot of energy to move the arm around, and can only move wheels relatively slowly—twice a second at best. This is fine for a quern, but it is nowhere near fast enough to power a lathe. Early Egyptian turners used two different methods to speed up the operation. Some used a bow to spin the axle, in the same way that early fire lighters spun their drills, but just like turning a handle this left just one arm free. Other turners employed a boy to provide the power. He wrapped a belt around the axle of the lathe and spun it by pulling the belt back and forth.

A better way of driving rotating devices, while keeping both hands free and minimising labour costs, is to use the feet. In classical times, potters mounted a platform at the base of their wheels, which they could propel around with a walking action. By medieval times, craftspeople had come up with a more complex system to power rotation: the treadle. One machine that used a simple example of this system was the medieval pole lathe. Pushing down on the treadle pulled down on a cord that was wrapped around the axle of the lathe, spinning the piece of wood. The other end of the cord, meanwhile, reached up to the end of a long springy pole and bent it downward. Releasing the treadle allowed the pole to straighten and the piece to spin backwards. Regularly pressing down on the treadle could therefore set up a reciprocating action in which energy was transferred between kinetic energy in the rotating lathe and elastic energy stored in the bent pole. An even better arrangement was to attach a rigid crank between the treadle and the wheel rather than a rope. Pushing the treadle rhythmically up and down could produce continuous rotation of the piece, especially if the lathe was equipped with a flywheel to store rotational kinetic energy.

Human-powered devices reached their apotheosis in the preindustrial age with the invention in tenth-century India of the spinning wheel. The big advance of this device was that rather than hanging from the thread, the spindle was held on a free-spinning axle. Early versions of the spinning wheel had a fairly simple design in which the spindle was driven by a belt that was wrapped around a large flywheel. The difference in size of the two axles acted as a simple gearing system; it was an effective way of producing rapid rotation of the spindle. The spinner rotated the flywheel using one hand, while using the other to tease out the fibres. But later versions had a far more sophisticated design. In the traditional spinning wheel, which is familiar to us from folk museums and from the illustrations in fairy tales, the flywheel is powered by a crank that is moved by a treadle, leaving both hands free to draw the thread. In the most advanced designs, the spindle is surrounded by a horseshoe-shaped structure, the flyer, which has a ring at the end of one of its arms, through which the thread passes before it reaches the spindle. Though they spin around the same pivot, the axles of the spindle and flyer are driven by separate belts from the flywheel and have slightly different radiuses. This means that they rotate at slightly different speeds, so as the thread is drawn into the machine it is automatically twisted before being wound onto the spindle; the spinner does not need to alternate between spinning the thread and reeling it onto the spindle, and so does not keep having to start and stop the wheel. The process can therefore be continuous, simultaneously speeding it up and making it more energy-efficient.

As we shall see later in the book, with a machine that rotates continuously, it is much easier to operate multiple identical shaping devices. This made it practicable to develop the huge factories of the Industrial

A traditional spinning wheel. The wheel is powered by a treadle, which drives the spindle and the horseshoe-shaped flyer at slightly different speeds to simultaneously twist the thread and wind it onto the spindle. The raw fibres are mounted on the distaff on the top left.

Revolution, with dramatic consequences for the modern world. But well before the eighteenth century and the advent of industrialization, the principle of the continuously spinning wheel had been exploited to build a vast range of machines that were capable of transforming energy efficiently from one form into another.

CHAPTER 9

Building Machinery

If there is one place in England that exudes old-world charm more than any other, it must be the Isle of Wight, an island off the south coast made up of pastoral fields dotted with chocolate-box villages of thatched cottages and surrounded by dinosaur-bearing chalk cliffs and sandy beaches. Within this bucolic refuge, the most charming place must be Carisbrooke Castle, a medieval fortress that was once used as a prison for Charles I. Its honey stone curtain walls surround the emerald lawns of the bailey and the stone tower of the keep, while looking out at a verdant countryside. And of all the attractions of the castle the most charming must be the sixteenth-century water treadwheel. Originally the wheel was powered by prisoners (though probably not His Royal Highness, who would have been above such menial duties), but in 1696 the role was taken over by animal power. Four times a day one of the castle's four resident donkeys is set to work walking inside the wheel, like an oversize hamster, to turn the axle that raises a bucket of water from the 160-feet-deep (50-metre-deep) well at the centre of the castle. Today's donkeys are well-fed and notoriously stubborn and lazy, so they are reluctant to work for more than a few seconds at a time, but watching them is a delightful spectacle. And it's also one that harks back to the invention and development of most of our power systems.

The donkey wheel in action at Carisbrooke Castle.

The treadwheel is one of the simplest of the rotating devices that people have devised to transform energy from one easily obtainable form to another more useful one. They were commonly used in Roman times, for instance, when they were powered by slaves, not only to raise water but also to move the arms of cranes. Of course, at Carisbrooke the donkey could simply pull the rope over a pulley to do the same thing, but the treadwheel has two advantages. It takes up far less space because the donkey stays in the same place as it spins the wheel. The process is also highly geared so that the donkey can raise a far greater weight of water, albeit more slowly, than if it were pulling a rope over a pulley. Indeed the gearing is much higher than could be achieved using a traditional crank to pull the bucket up, since the diameter of the wheel is far greater than the length of a hand crank. Pulling water up from a well might appear to be a rather niche activity, but as we shall see, the need to raise and pump water, and to extract energy from it, were crucial factors that drove the development of machinery and transformed the Old World even before the Industrial Revolution.

Since the advent of farming in the Middle East—a region that is subject to regular seasonal droughts—people have sought to extend the growing season and expand the area of land they could farm by the process of irrigation. They dug canals through the higher ground and lifted water into them so that it flowed to the adjacent fields. Irrigation is a surefire way of improving plant growth and increasing crop yields in arid regions, but it can be backbreaking work and uses a lot of energy. It is awkward to stoop down to the water to fill a bucket, and tiring to lift it, and much of the energy the farmer expends is wasted accelerating and decelerating the water and raising and lowering his body and the bucket itself in the process. By 2500 BC, the Mesopotamians had overcome some of these problems by developing one of the world's first machines, the shadoof. This is in fact just a simple pole that is hinged near one end to a solid framework. A bucket is tied via a rope onto the end of the long arm and a heavy counterweight at the end of the shorter arm. The shadoof acts as a simple lever, but it made, and continues to make, raising water much easier. For a start, the farmer no longer has to crouch down to the water, and rather than pulling upward to lift the water, thanks to the counterweight, he only has to pull downward on the long arm to lower the bucket to the water—a much easier operation. The counterweight performs the job of actually raising the water. Simple and easy to operate, shadoofs are still used, but they are relatively slow and still require quite a large muscular effort. Over the next few centuries, therefore, farmers came up with a whole range of new rotating devices that proved more efficient, and which enabled them to raise water without having to do any work themselves.

All these devices used the same principle: a rotating wheel to scoop up water, lift it, and release it higher up. They were the first devices to use gears, which were used to transmit the force provided by an animal walking around a circular treadmill to drive the machine. The first, and simplest, was the saqiyah, in which the wheel was fitted with buckets mounted on its rim; they picked up water as they rotated around the bottom of the wheel and released it as they rotated over the top. The saqiyah is more efficient than the shadoof because the buckets travel at a steady speed and the weight of the descending buckets cancels out the weight of the rising buckets, so the only energy needed is that used to raise the water.

Farmers later developed a further series of machines to raise larger volumes of water. The tympanum consisted of a circular drum that was separated into eight segmented compartments with an entrance hole at the front outer corner of each. As the drum rotated, water poured into each compartment through the hole and was carried upward until the segment reached the horizontal, when the water sloshed inward and poured out of another hole near the base of the segment. The tympanum had the disadvantage that it could only raise water by around a third of its diameter, while the sudden inward movement of the water within each segment also reduced its efficiency. Consequently, it was gradually replaced by two further devices. The scoop wheel, which was invented by the Egyptians, had many compartments that were curved into scoops that picked up water as the device rotated and moved it more gradually and gently upward and inward, before releasing it at the centre of the device, increasing efficiency from 48 to 55 percent. The principle was finally adapted by one of the greatest geniuses of the ancient world, Archimedes, to design his famous Archimedes' screw. This consists not of a drum but a long tube, which is separated by a continuous helical partition that runs at 45 degrees to its long axis. The tube is mounted at a small angle to the horizontal with the lower end in the water source, and is rotated so that water enters it and is

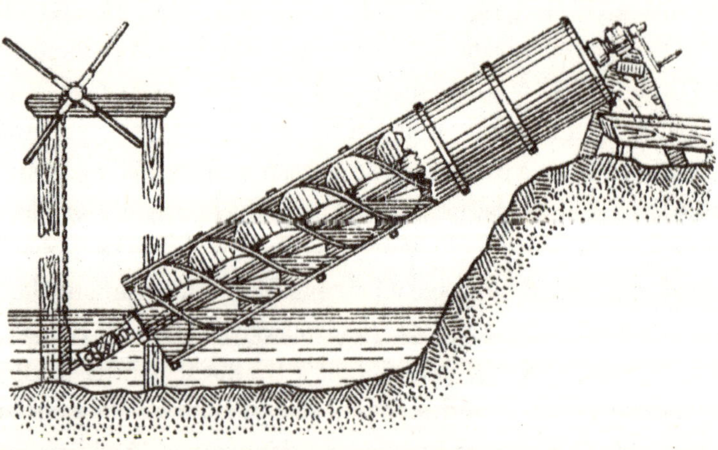

The most advanced of ancient irrigation devices, the Archimedes' screw. Rotating the device using the handle at the top draws water smoothly up the tube.

marshalled by the helical partition along, and gradually upward, to the top of the tube, where it is released in a continuous stream. Like the tympanum and scoop wheel, the lifting distance of the Archimedes' screw is limited, but it transfers water farther away from the water source, making it easier to build the gearing apparatus and the circular animal walk needed to power it. The Archimedes' screw is also more efficient—60 to 80 percent—than the tympanum and scoop wheel, since the water moves at a constant speed up the apparatus.

However efficient all these devices are, they still require an expenditure of muscular effort to power them, whether by man or beast, energy that they would have needed to recoup by eating some of the crops that the water enabled them to grow. In the fourth century BC, farmers in Egypt overcame this disadvantage with the invention of yet another machine, the noria. They realised that they could raise water from the Nile without any effort on their part simply by using the force of its flow. The noria is essentially composed of the drum of a saqiyah equipped with paddles that are mounted around the rim of the wheel. The water flowing past the device pushes the wheel around, thereby allowing the buckets to pick up water and raise it to the top of the wheel.

The first water-powered device, the noria. This example in Hama, Syria, uses the flow of the river hitting the paddles to turn the wheel and raise water picked up by buckets attached to the rim to the aqueduct.

In Europe, where rainfall is higher and the flow of rivers is more reliable, people were quick to realise the potential of using the flow of streams to power other processes than simply lifting water, the most obvious and pressing being the milling of grain. By classical times they were already using rotary querns, so the idea of powering a quern by harnessing it to some form of waterwheel would have been an intuitive one. The Romans developed several types of waterwheels to mill grain, their devices being described by the first-century BC engineering writer Vitruvius. Probably the first devices they built were undershot wheels, which were essentially barely modified norias. They were coupled to power quernstones using the same sort of gear system that had already been developed to power saqiyahs and Archimedes' screws, but in reverse; gear pegs on the horizontal axis waterwheel interlocked with pegs on a vertical axis gear that drove the upper quernstone.

Undershot wheels suffer from two disadvantages: first, they only work well when the water level exactly covers the blades, and become less effective if the river level rises or falls; moreover, some energy is lost in the transmission of the movement from horizontal axis to vertical axis. The first difficulty was overcome by the development of the overshot wheel. Part of the stream was diverted from the main flow along a millrace to the wheel. It was directed into buckets at the top of the wheel, and the weight of the water, rather than its momentum, drove the wheel around as the wheel rotated and the full buckets of water fell. Overshot wheels quickly became popular. The most impressive remains of Roman milling are found at Barbegal, near Arles, France, where an aqueduct supplied water to the tops of two flights of eight overshot waterwheels, making sixteen wheels in all. It has been estimated that this installation must have been capable of grinding over four tons of flour per day, enough to feed ten thousand people.

Another way of powering a millstone, one that did not need any gearing, was to use a vertical axis waterwheel. In this design, part of the stream could be diverted not into a millstream, as for overshot wheels, but into a sloping pipe, which converted the gravitational potential energy of the water into a rapid flow. A nozzle at the end of the pipe directed the water against angled blades that were arranged

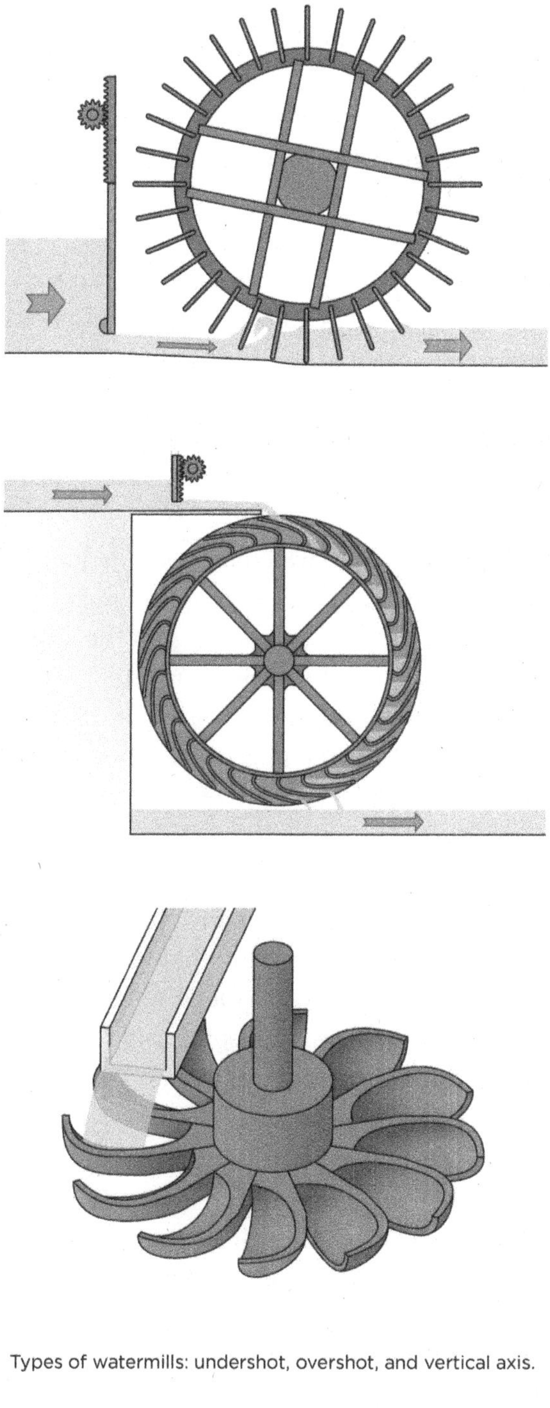

Types of watermills: undershot, overshot, and vertical axis.

around the edge of a horizontal wheel that was on the same axle as the quernstone. The impulse pushed the apparatus around. Vertical axis wheels worked better in hilly districts with fast-flowing streams, but since these areas are less densely populated, vertical axis wheels were never used on a large scale. In contrast, horizontal axis waterwheels were best suited to large slow-flowing rivers, and as we shall see, they later came to dominate industry.

Of course, not everywhere has a convenient source of energy in the form of flowing water, but waterwheels must have acted as the inspiration for people intent on extracting energy from another form of fluid flow: the wind. The first windmills borrowed the principle of the undershot waterwheel. Developed in Persia in seventh century BC, the panemone is a drag-driven device, with seven or eight sails of cloth or wood mounted around a vertical wooden shaft. The mills sit within gaps in a massive stone wall that is oriented in a northwest to southeast orientation so that they can intercept the prevailing northeasterly winds of this subtropical region. And the shaft is mounted to one side of the gap, so that on the open side the sails are driven downwind, while the returning sails on the other side are sheltered by the projecting wall. Because the sails intercept only a small fraction of the total area of wall, and because of the high drag on the sails, panemones are very inefficient devices, but they are still used in parts of Iran.

In Europe, where both the speed and the direction of the wind are much more variable, people found that a more practicable way of extracting wind energy was to use the principle of the vertical axis waterwheel. The wind could be intercepted using angled sails of cloth or wood, mounted on a horizontal axis wheel. This arrangement had the disadvantage that it needed gearing to transmit the movement downward into a vertical shaft, but it meant that the sails could intercept a far greater area of wind and could be readily turned towards the wind as it changed direction. Having been introduced into Europe in the tenth century from the Middle East, people quickly developed windmills, particularly in the windier parts of Northwest Europe. The first European windmills were small square wooden structures

Windmills ancient and more modern. In the panemone (*top*), the wind travels through slots in the mud-brick wall, pushing the vertical vanes around using drag. In the tower mill (*bottom*), the wind is intercepted by the sails that are driven around by lift, like a propeller. The fantail at the back keeps the sails turned into the wind by rotating the cupola.

called post-mills, whose whole structure was built around a central wooden pillar. The whole building could be wheeled around the pole to face the wind using a tailpole at the rear. Later, larger mills also had a more permanent stone structure. In smock mills, the stone ended some third of the way up, and the wooden upper part of the building, which was usually octagonal, still rotated. In the later tower mills, which reached their apogee in the Victorian era, the whole cylindrical structure was made of stone apart from a hemispherical wooden cupola that housed the blades. The cupola was mounted on cogged wheels that were linked to a smaller set of blades, the fantail, that projected on the opposite side from the arms of the mill. This acted as a self-orienting device; if the wind veered away, it would cause the fantail to rotate, which would drive the cupola around, so that the sails faced into the wind again. In this device, therefore, spin was built into many of its components.

Windmills were cheap to build, but costly to run, so they were generally built only in areas where water mills were impractical, which included chalk hills, where there were no surface streams, and marshland where there was no flowing water. Consequently, one of the areas with the highest concentration in Britain must be the region around Hull, where I live. The chalkland of the Yorkshire Wolds is still dotted with the huge black tower mills that used to grind the grain grown on the fertile soils, while next to the dazzling twentieth-century icon of the Humber Bridge, there lies the remains, now restored, of a wind-powered whiting mill that ground up the chalk quarried at the site to make lime mortar. And since the water that emerges from the base of the Wolds appears barely above sea level, the rivers that drained the flat landscapes of Foulness and Holderness were unable to power any water mills. The city of Hull, therefore, used to be surrounded by large numbers of windmills that ground the grain that was shipped down to the Humber estuary, and the linseed to produce the oil that was used to make paint.

But it was in the Netherlands where windmills proved to be the most important and where they reached the height of their development during the Dutch "golden age" in the sixteenth and seventeenth centuries. As the Dutch excavated the huge areas of low peat that lay in the Holland area, and floated the dried peat away to fire the bricks and pantiles to build their beautiful cities and power their industry, they needed some way to pump the water off the flooded land and convert it into farmland. Since the land was so flat, there was no way they could use waterwheels. The solution was to use lines of windmills to power scoop wheels that lifted water and released it on the seaward side of their dikes. To increase the power and efficiency of the mills, they gradually developed the design of their sails, separating each sail into a large number of slats that could act just like the flaps of modern aircraft to maximise the energy they extracted from the wind.

One of the few downsides of water and windmills is that the power they produce comes in the form of rotational motion. This is fine for some processes, such as grinding grain and scooping up water, but most of the heavy industry of the ancient and premodern world demanded a quite different motion: reciprocating. In the textile industry, for instance, woollen cloth needed to be hammered with mallets to thicken and waterproof it, a process known as fulling. Ironmasters needed to ventilate their furnaces with bellows to smelt their iron and to hammer the iron they produced to remove slag and convert it into bars. And sawyers needed to swing their saws back and forth to cut tree trunks into planks. To mechanise these processes, early engineers needed to devise techniques to convert rotational motion back to reciprocating motion.

The first solution to be devised was the cam, which was simply a peg mounted onto an extension of the axle of a wheel, called the camshaft. For instance, in fulling mills and forges, the pegs lowered the base of the handles of huge hammers, which were hinged near their base, lifting and then releasing the heads so that they fell under their own weight onto the cloth or iron bloom. The builders of the first sawmills invented a more sophisticated method to power their saws, reversing the operation of the crank that had been developed to power spinning wheels. The rotating wheel drove a crankshaft that at its upper end was joined to one end of the saw, which was constrained to travel between vertical slots so that it could only slide up and down, while a mechanism moved a log horizontally along the apparatus, allowing it to be sawn into long planks.

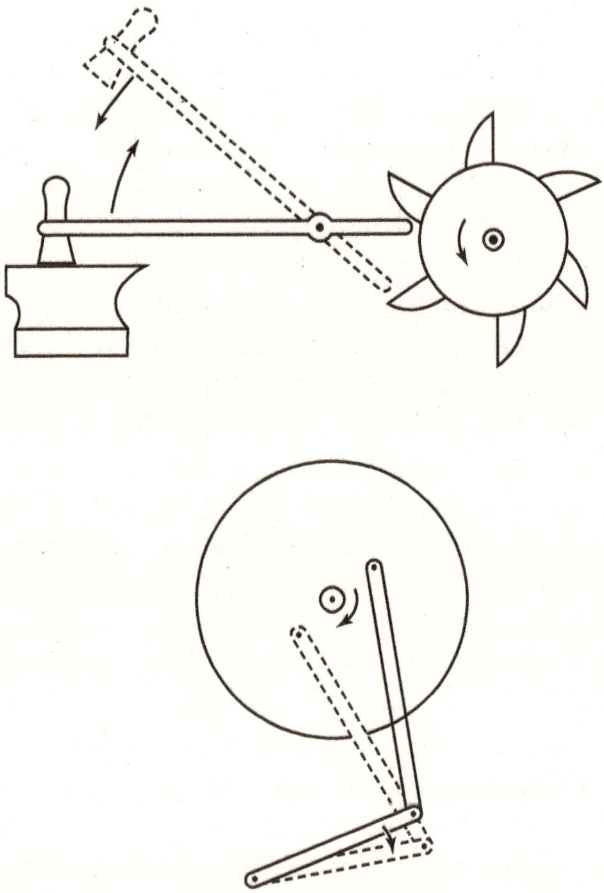

The two main methods for converting rotational motion into reciprocating motion. In the cam (*top*), projections from the axle slide over the handle of a hammer or the arms of a bellow. In the crank (*bottom*), a series of joints converts the motion. Note that the crank can also be used in reverse to convert reciprocating motion into rotational motion.

Other solutions were needed to overcome a final drawback of all the water-raising equipment we looked at in the first part of this chapter: their inability to raise water above the top of the apparatus. Over the last thousand years, engineers developed several different methods to achieve this, in all of which a rotating wheel was used to power a reciprocating piston pump. The famous thirteenth-century Arabian scientist and inventor Al-Jazari came up with a series of beautiful geared devices. In the fifth of his ingenious designs, a waterwheel drove a vertical gearwheel on the side of which was a peg. The peg inserted into the groove of a slot rod, so as the wheel rotated the rod was moved from side to side, pushing and pulling on the pistons of two cylinders that were mounted horizontally on either side. As one piston was on its delivery stroke, the other was on its suction stroke so that the device acted as a double-action reciprocating pump that could raise water hundreds of feet.

In Europe, engineers used the principle of the crankshaft to do the same thing. Rather than push a rod along its length, as happened in sawmills, the crank was jointed at right angles to a rocking arm or beam so that rotating the wheel moved the beam up and down. The other side of the beam was in turn joined to the push rod of a reciprocating pump so that its movement alternately emptied its cylinder to push water uphill and allowed the cylinder to refill with water. A beautiful water-powered reciprocating pump is on view at Paxton House in the Scottish Borders, where it was used to raise water 80 feet (25 metres), from the Paxton Burn to the great house. By the 1850s, American inventors had also developed wind-powered reciprocating pumps that could be used by settlers to raise groundwater to the surface for irrigation and for drinking water. The axle of the multi-fanned wheels, which are so familiar to us from Western films, was geared to a larger wheel to slow down the motion. In turn the large wheel drove a crank up and down, moving the piston of the pump in and out of the cylinder, lifting water.

By the end of the premodern period, therefore, the power of water was being widely exploited for industry, especially in Northern Europe. In England, for instance, rivers and streams could be dammed every few hundred yards and industry covered the banks of what we nowadays often think of as unspoiled "natural" stretches of water. The wool-producing areas, such as the Cotswolds and Pennines, contained large numbers of fulling mills, and the heavily wooded Weald was dotted with ironworks containing mills that drove hammers and bellows. And where there were no rivers the power was obtained from the wind. Far from being handmade, even well back in the Middle Ages, many consumer items were made using surprisingly sophisticated machines that obtained power from renewable sources, and which converted it from linear to rotary motion and back again.

CHAPTER 10

The Industrial Revolution

A visit to an industrial museum is invariably an assault on the senses. Whether it be a working textile mill, a foundry, a pumping station, or a steam railway, the visitor is overwhelmed by the smells of coal and oil; deafened by the clanking of machinery, the whirring of gears, and the hiss of steam; and mesmerised by the relentless pumping of pistons, the march of belts and chains, and the spin of wheels. These places evoke feelings of irresistible power as the machinery rolls inexorably onward. So the term "Industrial Revolution" seems entirely appropriate to describe the period from 1700 to 1850 when all this mechanical bandwagon was set in motion. The machinery you see in these museums seems to have little in common with the stately tread of donkeys, the slow rumble of a water mill, or the almost silent sweep of a windmill's sails.

So it is tempting to think that the progress of the eighteenth century must have owed much to the scientific revolution that had begun in the seventeenth century, and that had culminated in the triumph of Newtonian mechanics. But the fact is that the Industrial Revolution was driven by mostly uneducated practical men and women, just as earlier technologies had been developed by thread-making housewives, fire-lighting hunters, and top-spinning children. Driven by the desire to speed up the production of goods, they simply applied established principles. They refined and combined rotating devices that already existed, but used them in novel and unexpected ways to build a new technological world. And as we shall see, each invention was built on previous ones, or borrowed from inventions in other fields to drive an unprecedented process of change that fed on itself to transform both the world's economy and its environment.

One of the first major technological advances that kick-started the Industrial Revolution occurred, as it had in prehistory and the pre-modern age, in the textile industry. In this case, however, it involved an improvement not in spinning, but in a process that was based on reciprocating motion—weaving. In a traditional treadle loom, the main horizontal frame holds two alternating sets of warp threads that are threaded at the centre through holes in two vertical lines of wires, the heddles. These are held within two vertical frames and are linked to the treadles so that pushing down on the right treadle will lift one frame and lower the other, while pushing down on the left treadle does the reverse. Between these actions the weaver leans forward and passes a bobbin with a weft thread on it—the shuttle—from one side of the loom to the other. The weaver finally grabs yet another vertical frame—the batten—which holds vertical lines of reeds that are set between each warp thread, and pulls it towards them, to compress the weft fibre into the growing piece of cloth. After pressing the other treadle, the frames reverse their position so that the new weft fibre crisscrosses the warp fibres in the opposite direction, producing strong connections between them and building up the cloth.

Apart from being extremely bad for a weaver's posture—weavers were notoriously round-shouldered—leaning forward to pass the shuttle back and forth was a slow business. Unless the weaver hired an assistant, it also limited the width of the cloth, or "piece," they could produce to a couple of feet. The solution to overcome these limitations—to make broader pieces of cloth, to weave faster, and to improve weavers' postures—was to incorporate a smooth track across the base of the batten. The redesigned shuttle, which was now bullet-shaped and equipped with pairs of metal wheels front and rear, could roll back and forth along this track, taking the thread with it across the cloth, before being caught in a leather box at the end of the batten. Each box was attached to strings that the weaver simply tugged to propel the shuttle on its return journey. The flying shuttle, as it soon became known, was first invented in Languedoc, France, in 1732, but the shuttles were destroyed by state cloth inspectors, who were determined to keep the status quo—and their salaries. Fortunately, the idea made its way to England, where it was patented in 1733 by a Lancashire weaver, John Kay, who had already invented a new way of using wires rather than reeds in the battens of looms. Kay always called the device a "wheeled shuttle," and indeed it was probably the first application of the moving wheel that was totally successful—largely because the shuttle rolled along the perfectly smooth surface of the track. The flying shuttle doubled the productivity of weavers, and allowed a single weaver to produce broad cloth, but Kay was not the first or last innovator who was destined not to fully reap the rewards of his ingenuity. Manufacturers in England deliberately infringed his patents and banded together to frustrate his lawsuits, and the French government, to whom he later turned, proved unwilling to keep paying the pension it had awarded him.

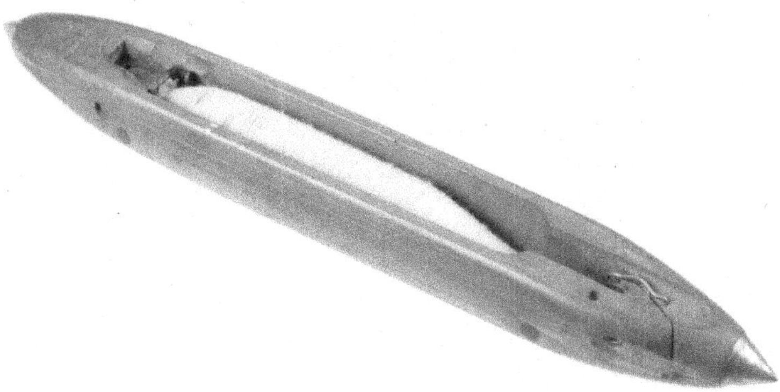

The flying shuttle. Possibly the world's first wholly successful wheeled vehicle. The streamlined bobbin holder runs on wide wheels along the track of the batten.

The flying shuttle also created a problem, since it disrupted the balance between spinners and weavers. The textile industry now needed a way of speeding up spinning, and in the 1760s, two new machines were invented to do just that. The best known—perhaps because of its charming name—was the spinning jenny ("jenny" being a diminutive of the word *engine*). Invented by another Lancastrian, the weaver James Hargreaves, this was essentially a spinning frame in which the operator could spin eight or more threads simultaneously by turning the handle of its single wheel. The machine worked reasonably well, but the thread it produced was uneven and weak, the problem being the way in which the cotton fibres were drawn. In a traditional wheel, the spinner stretched them gently between the fingers of her free hand; it was a matter demanding quite some skill. In the spinning jenny, this stage was carried out simultaneously in all the threads by pulling the cotton roving along between two bars, an arrangement that was necessarily less precise. Consequently, thread produced by a spinning jenny was less consistent and weaker and could only be used for the weft threads in a loom. Although two hundred thousand spindles were making weft threads in this way as early as 1788, when Hargreaves died, another method was clearly still needed to produce warp thread. And while it was the wheel that had sped up weaving, it was rollers that sped up the production of warp thread.

In the machine patented by the Preston wigmaker Richard Arkwright (another Lancastrian), and built for him by the clockmaker John Kay (yet another Lancastrian, but no relation to the first John Kay), the roving was passed through three sets of rollers that rotated at increasing speeds, and so gently stretched it. This mimicked the behaviour of a spinner's fingers and produced even yarn that it then twisted into strong threads. The only problem was that because of the friction in the rollers, the machine consumed too much energy to be powered by a single person. Arkwright experimented with horsepower in his first mill in Nottingham, but this was still not enough, and he turned to waterpower, hence the popular name of his machine—the water frame. In 1771 he moved to the weaving village of Cromford, Derbyshire, and built the world's first factory on the banks of the River

The spinning jenny was basically just a spinning wheel that spun multiple threads at the same time.

Derwent, the world's first works' housing development in the village for his workers, and the world's first company pub, the Greyhound Hotel. All are still there to this day, and well worth a visit; the village combines industrial archaeology, with the charming rural surroundings of the Peak District, and even an excellent secondhand bookshop.

Not content with revolutionising spinning, Arkwright also patented a novel carding machine. It converted raw cotton wool into rovings for his spinning machine, once again using a series of rollers. However, in this case the rollers were covered in mats of hooked teeth so that as they sheared past each other the teeth teased out the fibres and removed dirt to produce a fine stream of perfectly aligned cotton fibres. The American industrialist Eli Whitney used a similar principle to remove the seeds from raw cotton. The cotton fibres had traditionally been cleaned using pads covered with wire teeth that were pulled across each other, in a process very similar to carding. In India, this process had been speeded up by using rollers driven by a

Arkwright's water frame. The cotton was gently stretched by passing through three sets of rollers toward the top of the machine, before being twisted by the rotating bobbins toward the bottom.

hand-powered worm gear. Whitney added waterpower to the process, producing his famous cotton gin ("gin" being yet another diminutive for *engine*), which speeded up the process and reduced labour costs a further tenfold. It was the cotton gin, more than any other invention, that allowed the cotton plantations of the Southern states to become profitable and expand in the first half of the nineteenth century, increasing their dependence on slavery.

The water frame was not the last word in spinning machinery. It was swiftly followed in 1785 by a machine that combined the action of the spinning jenny and the water frame, and that was consequently called the mule. Designed by Samuel Crompton (another Lancastrian, unsurprisingly), these magnificent machines operate in two mesmerising stages, just like a hand spinner. In the first stage, rollers draw out the rovings as the carriage on the side of the machine moves out on wheels, and the thread is twisted. In the second stage, the carriage returns, and the machine winds the spun thread onto the spindle. Mules dominated the Lancashire spinning industry well into the twentieth century, producing thread of unsurpassed quality. They were only made redundant in the last fifty years by ring spinning machines, direct descendants of Arkwright's water frame, which underwent further development in the cotton mills of Rhode Island, the heart of America's fast-expanding cotton industry.

Arkwright's mill at Cromford proved to be the prototype for factories throughout the world in the eighteenth and nineteenth centuries. They were powered by huge overshot waterwheels, built according to the designs of the engineer John Smeaton, who we will meet in the next chapter, and they were tall, multistory buildings. This might seem odd, considering the weight of the machines they had to house, but in a factory powered by a single wheel, the machines need to be as close to it as possible. In a multistory mill, the waterwheel could be geared to a vertical power shaft that rose up through the floors, and on each floor it could be geared to drive horizontal shafts. The energy was finally run off these shafts by linking them to each machine using simple belt drives.

Later, mill owners also moved weaving machinery into their factories, where they could drive their flying shuttles using cams that flicked levers to accelerate them back and forth and to move the frames up and down. With all their reciprocating movements, power looms made a terrible noise that deafened the workers, who had to communicate through hand gestures. The great majority of the British cotton mills were set up, as you might have guessed, in Lancashire. There were, of course, plenty of ingenious engineers to improve the technology in the county, but Lancashire also had geographical advantages. The predominant southwesterly winds that blow over the county bring damp Atlantic air that releases its moisture onto the western slopes of the Pennine Hills. Not only does this create a permanent damp atmosphere, which helped the cotton fibres stick together, reducing spinning costs by a quarter, but it also filled the streams and rivers that flowed down the slopes,

In a traditional water- or steam-powered textile mill, the power was transmitted to the machines using a complex system of cogs, shafts, and belts.

providing soft water to wash the cotton and dye the cloth. The rivers also powered the waterwheels of what became the mill towns of Oldham, Bolton, Bury, and Rochdale, before joining up to form the River Irwell and transporting finished cotton goods into Manchester, the world's first industrial city with the world's largest cotton exchange. Finally, the Irwell joined up downstream with other rivers to form the Mersey, down which the finished cotton was floated to Liverpool and the sea.

But it doesn't rain all the time, even in Lancashire, though to its inhabitants it may sometimes feel like it does, and mill owners became frustrated by the limited capacity and unreliability of waterpower. The obvious solution was to augment, or replace, their waterwheels with the new power source of the Industrial Revolution, the steam engine, which by the 1770s had been improved by James Watt's invention of the separate condenser. Steam engines could operate around the clock whatever the weather, but they had a disadvantage when it came to driving machinery since they produced reciprocating motion. The early beam engines had a vertical cylinder with a piston that drove a horizontal pivoted beam up and down, raising and lowering the cylinder of a reciprocating pump. The first steam engines to be introduced into mills were therefore simply used to pump water back up to the millpond during times of drought, but this was clearly an inefficient use of their energy. To power machinery directly instead, the vertical motions had to be converted into rotation. The obvious way to do this was simply to attach a crank to the beam and use it to rotate a flywheel, reversing the way in which waterwheels had long been used to pump water. However, James Watt was unwilling to do this for two reasons.

First, because his engines produced power on both the upstroke and downstroke, he worried that the vertical movement of the piston could not safely be linked to the beam, which moved slightly in and out as it moved away from the horizontal. To overcome this, Watt came up with the invention he was most proud of—the parallel action. Instead of coupling his pistons to the beam, he joined it via two sets of parallel rods that would swing in and out automatically as the beam rose and fell.

Second, Watt was concerned that the action of a flywheel would interfere with the intermittent power stroke of his engines. He need not have worried. His rival, James Pickard, not only equipped his engines with cranks but also patented the idea, and showed that the smoothing action of the flywheel actually improved the performance of steam engines. Nettled, and unwilling to pay a license fee, Watt determined to produce an alternative to the crank. The only design that worked was the one put forward by his employee, William Murdoch: the sun and planet gear. Instead of being attached directly to the flywheel, the crank ended in a fixed cogwheel, which was held by a bar so that it interlocked with a similar cog attached to the flywheel. As the beam rose and fell, the "planet" gear rotated around the "sun" gear, causing it to rotate at twice the frequency of the engine. This arrangement worked, but it added another degree of complexity and more friction to the system, so when Pickard's patent ran out, Watt reverted to using the simple crank.

Watt also came up with a mechanism to regulate the speed of his engines. In his centrifugal governor he adopted a device that had been invented a century before by the Dutchman Christiaan Huygens to control the distance between millstones in the country's windmills. This consisted of two balls that were attached to a shaft that was driven by the engine. As the speed of the shaft increased, centrifugal force pulled the balls outward and upward, closing a valve and thus reducing the power input to the engine and slowing it down again. Centrifugal governors continued to be used throughout the nineteenth century; even today the same principle is used to control the speed of snowmobiles, and to regulate the striking train in repeating watches.

Watt's rotative steam engine. Note the sun and planet gear to turn the flywheel on the left, the parallel action connecting the cylinder's piston rod to the beam at the top right, and the centrifugal governor with its two balls upper centre.

It soon became clear to engine designers that steam engines could work just as well when their piston was oriented horizontally as vertically, allowing them to be fitted into smaller buildings with lower ceilings. This arrangement also simplified the coupling of the piston to the flywheel. Engineers could constrain the piston rod to move back and forth along a lubricated trackway and link its far end to a crank. This arrangement was, of course, simply the reverse of the one that had long been used by stonemasons, who had used a rotating waterwheel to drive a horizontal saw. With a power source that could be located almost anywhere, mills could be concentrated together in large industrial towns. In Lancashire, for instance, mills were increasingly set up in the flat environs of Manchester, to which raw materials could be more easily imported and cloth exported along the developing canal network. The downside was the massive air pollution caused by burning sulfur-rich coal. Black smoke hung over Manchester like a blanket, and everywhere they were used, steam engines caused pollution that was to scar the planet for centuries.

While Watt and his rivals repurposed old ideas to convert the action of reciprocating steam engines into rotation and so power their factories, ironmasters did the same to deal with their new material—wrought iron. In 1783, the Lancashire ironmaster Henry Cort developed a new way to produce large amounts of a novel form of iron that had the same toughness as the bar iron that blacksmiths had traditionally produced using the lengthy bloom process. Iron was toughened by alternately heating and hammering it. In Cort's "puddling" process, the operator melted together large amounts of cast iron, along with iron oxide and slag to burn off excess carbon and incorporate toughening slag fibres. The process speeded up iron production fiftyfold. The only problem was how to shape the huge mass of spongy molten metal that a puddling oven produced; hammering it out would take huge amounts of time and energy, negating the advantages of the puddling process. Cort's solution was to use a series of steam-powered rollers, like giant mangles, that could flatten the iron into useful rods and plates. Hot rolling quickly became the primary method of shaping wrought iron, and it remains the main way of shaping steel components to this day.

Wrought iron proved perfect for structural work, since it could withstand both compression and tension and enabled industry to expand further. Mill owners such as Derby's Joseph Strutt used rolled wrought-iron beams with an I-shaped cross section to support the floors of ever-larger iron-framed factories. Even today similar beams, rolled steel joists (or RSJs as they are known in the industry) dominate engineering, forming the frames of skyscrapers and bridges alike. Mill owners could also power larger mills using new high-pressure steam engines, which became feasible with the advent of wrought iron. High-pressure boilers could be made simply by riveting wrought-iron plates together. Using high-pressure steam, steam engines became not only more powerful but also more efficient, allowing yet more processes to be mechanised. Hence, in the early years of the nineteenth century, engineers such as Eli Whitney in America, and Henry Maudsley and Joseph Whitworth in England set up the first steam-powered machine shops. Together, these engineers developed the machine tools that were to build the modern world, from the first screw-cutting lathes to the rotary tool that more than any other allowed engineers to make precision instruments, the milling machine. These consist of a fast-rotating head that can be moved precisely in all directions and cut perfectly flat surfaces.

All this rotating technology came together in the first half of the nineteenth century to transform transport and shrink the world, both on land and at sea. With high-pressure wrought-iron boilers, steam engines could at last be built small and light enough to propel themselves. The first steam locomotives—such as the Cornish engineer Richard Trevithick's "Catch Me Who Can" of 1801—travelled along roads, but encountered the same problems as horse-drawn carriages. The roads were simply too bumpy for them in the summer, and too soft and muddy in the winter. The solution was to adopt the technology that colliery owners had used for two hundred years to transport their coal from their mines to the sea: to run their vehicles along rails. The only problem was that steam engines were still too heavy—they wore down wooden rails and broke brittle cast-iron ones. Railway engineers had to use rolled wrought-iron rails. From 1830 and the opening of the first passenger line, the Liverpool and Manchester Railway, right up to 1850, railway fever broke out. Soon not only Britain, but all of Europe and the fast-developing United States had been crisscrossed by thousands of miles of railway tracks, transporting record numbers of people and weights of goods at unheard of speeds of over thirty miles per hour across the continents.

Meanwhile, naval architects combined high-pressure steam engines and rolled wrought-iron plates to develop an even more important mode of transport, the iron steamship. They built the hulls of steamships by riveting wrought-iron plates together and reinforced them with wrought-iron bulkheads and rolled wrought-iron joists. The ships were powered by huge high-pressure steam engines. And the problem of how to propel early steamships was solved by once again reversing an old technology: in this case the undershot waterwheel. Huge paddle wheels set on either side of the ship accelerated water backwards or forward. As anyone who has ever hired a paddleboat or pedalo knows, paddle wheels have both advantages and disadvantages. On the plus side they are very easy to turn, since you can spin the two paddles in opposite directions. On the downside, the paddles churn up the water, wasting energy, so they are very inefficient. And depending on how loaded the pedalo or paddle steamer is, and hence how deep in the water it floats, the paddles can become even more inefficient. As early as the 1820s, engineers experimented with other mechanisms to propel ships, employing such technologies as Archimedes' screws. They soon found that the best design was the propeller, a screw with just one set of twisted blades, rather like the sail of a windmill. The superiority of the screw propeller over the paddle wheel was comprehensively proved in 1843 to 1845, when the British Admiralty ran a series of trials, pitting screw-driven ships against ones with paddle wheels. The most famous test occurred in March 1845, when the screw-driven HMS *Rattler* beat the paddle steamer HMS *Alecto* in several races, finally besting it in a tug-of-war contest. The trial ended with *Rattler* towing *Alecto* backwards at a rate of two knots. Soon all steamships were fitted with propellers, apart from a few, such as the famous Mississippi paddle steamers, which plied their trade in shallow waters where a screw could easily be damaged by hitting the riverbed.

The screw-propelled HMS *Rattler* (*left*) towing the paddle steamer HMS *Alecto* (*right*) backward. Note the direction of the smoke and flags!

By 1850, the world had been transformed forever by a whole series of innovations that had applied simple established principles of rotating bodies to produce novel machines and forms of transport. The innovations had transformed the productivity of the textile and metalworking industries and revolutionised transport, starting the process of globalization that continues to this day. And it was all done by practical engineers with little if any involvement of scientists or mathematicians. It was only later, in the nineteenth century, that scientific experiment and mathematical analysis were used to any effect to improve the power output and efficiency of machines and help build our modern technological world.

CHAPTER 11

Turbines, Pumps, and Generators

If there is one island off the British mainland that rivals the Isle of Wight in charm, and trumps it in historical and engineering interest, it must be the Isle of Man. Before it became a Victorian seaside resort for the cotton workers of Lancashire, this island in the middle of the Irish Sea was battled over for centuries between the English, Scots, and Vikings, and was for a time an independent fiefdom with its own parliament, Tynwald, which sits to this day. The rugged landscape of mountains, cliffs, and wooded glens is studded with ancient archaeological sites and medieval castles. And to cap it all is a heritage of nineteenth-century engineering: a mountain railway that uses a rack-and-pinion system to help it climb to the top of the island's highest peak, Snaefell; a narrow gauge steam railway; and both electric and horse-drawn trams. But the most impressive sight must be the great Laxey Wheel; at 72.5 feet (22 metres) in diameter and 6 feet (1.8 metres) wide, it is the largest working waterwheel in the world. Adorned with the symbol of the island, the three legs of Man, it sits out in the open, supported by a massive masonry and iron framework. It is fed by water piped from the surrounding hills, and is linked via a crank-and-beam arrangement to pumps that used to drain the Laxey lead mine.

The great Laxey Wheel, the largest working waterwheel in the world.

The Laxey Wheel, built in 1855, is just the most spectacular of the many power plants from the nineteenth century that enthusiasts have preserved and which are nowadays popular tourist attractions. They range from giant waterwheels to the massive beam engines that used to pump water around expanding cities and power Mississippi paddle steamers, the horizontal engines that powered factories and forges, and the huge steam locomotives that still run on heritage railways. Who is not swept away by their massive size, or awed by the power of their inexorable rhythm, the way they seem to move like giant beasts? Who is not apt to rhapsodise about the engineering genius of our Victorian forebears and lament the passing of the great age of steam?

But the prosaic fact is that, impressive as they are, these power plants have been superseded by better ones. For all their pomp, they suffer from many disadvantages. For a start, they are all hugely bulky and expensive. To build the Laxey Wheel, engineers had to divert watercourses, blast through rocks to install the pipework and pump, and construct a huge frame to hold the wheel itself. Beam engines needed to be built on massive foundations and filled buildings that resembled the naves of cathedrals. And despite appearances, Victorian power plants have relatively small power outputs, largely because they move so slowly. The Laxey Wheel rotates just three times a minute and produces just 180 horsepower (around 130 kilowatts). This is about the same as the 400-pound (180-kilogram) rotary engine that powered the Sopwith Camel we met in the prologue, which spun at 1,250 rpm. The Laxey Wheel could not rotate faster, because if it did water would be spun out of its buckets by centrifugal forces. Steam engines were also limited in power output because if they moved rapidly, the reciprocating motion of their pistons would set up huge inertial forces that would shake the machine to pieces. This was a particular problem for steamships, which meant that most were underpowered. It proved the main reason behind the failure of Isambard Kingdom Brunel's record-breaking ship of 1858, the SS *Great Eastern*, which, at eighteen thousand tons, was over three times as big as the weight of any other ship at the time. Its four huge steam engines struggled to keep moving and proved highly temperamental.

A final disadvantage of early nineteenth-century engineering was that each mill or pumping house had to have its own engine or wheel, since the mechanical power they produced could only be transmitted a short distance; each was in effect an isolated "off grid" unit. They were heavy and difficult to move, so in between fixed power plants, engineers still had to rely on muscle power, whether that was from draft animals or people. The canals and railways of the nineteenth century, for instance, were built by huge numbers of itinerant workers of "navvies" wielding spades, pickaxes, and hammers. To expand industry further, it was clearly necessary to improve the efficiency of power plants: to make their operation smoother, to make them smaller, to make them move faster, and to find better ways of transmitting the energy they produced. And surprisingly, the first successful attempts to achieve these aims were performed using the oldest of all technologies: waterpower.

Engineers had been debating about the relative merits of overshot and undershot waterwheels since Roman times, without coming to any firm resolution. And it was not until the middle of the eighteenth century that anyone attempted to tackle the question experimentally. The young British engineer John Smeaton devised a series of model tests to compare the performance of various designs of waterwheels, and he published his results with the Royal Society in 1757. Smeaton went on to have a brilliant all-around engineering career, building the world's first successful lighthouse on offshore rocks, the Eddystone Lighthouse; constructing canals, bridges, harbours, and roads all around Britain; and draining large areas of Eastern England. He therefore approached the problem of waterwheel design in a practical way. He made a series of scale models and used them to determine the amount of useful work each type of wheel could produce, by measuring how far each mill design could raise a set of weights for a given fall of water. He was also able to account for the friction in the machinery and came up with unequivocal results; overshot wheels were on average twice as efficient as undershot ones. Satisfied, Smeaton went on to produce practical recommendations for the design of water mills of different sizes, a major boon as water-powered textile mills were being set up throughout Britain in the second half of the eighteenth century.

However, Smeaton was less clear about the reasons for the differences in performance. He suspected that in an undershot wheel, power was lost in the turbulent collision between the water and the paddles, and he recommended that in overshot wheels, water should enter the buckets as gently as possible. But since at the time the concept of energy had not even been developed, he understandably went no further. Consequently, it was only fifty years later that the next advances in waterpower technology were made, and not in Britain, which was dominated by practical engineers, but in its neighbour, France.

In the eighteenth century, progress in mechanics was largely confined to France, where their mathematicians extended Newton's work on motion to rotation, and in the early nineteenth century it was again French mathematicians such as Gaspard-Gustave de Coriolis and Nicolas Léonard Sadi Carnot who started to define the concepts of kinetic and potential energy. The first practical benefits of this new understanding were supplied by a hydraulic engineer, Jean-Victor Poncelet. Like Smeaton, Poncelet realised that the energy loss in undershot wheels was due to the turbulence produced when water slammed against the flat paddles. He went on to reason that they could be made more efficient if the paddles slowed the water down more gradually. Poncelet's solution was to mount large numbers of gently curved metal blades onto the wheel. Water would run across the leading edge and rise smoothly up the blade before falling back down and running out of it again. Since the blade itself was moving forward, the water flowing backwards out of the blade would now have a negligible net velocity, so all of its kinetic energy would have been harnessed. The Poncelet wheel, as it became known, was an engineering triumph. Capable of extracting 73 percent of the energy of the incoming water, light, cheap to build, and far easier to install than an overshot wheel, it could be used on a wide range of watercourses; hundreds were built throughout France. However, Poncelet realised that his design had limitations: it still demanded engineering works to set up the wheel; and it had the disadvantage that water exited the blades from the same place as it entered, limiting its maximum speed. Nevertheless, Poncelet's success persuaded the French government that waterpower could be the key to industrialization—a way for a country with few easily exploitable coal reserves to catch up with its rival across the channel, Great Britain. The best way of doing this seemed to be to improve the design of the vertical axis waterwheels that were already in use in mountainous areas of the country where Poncelet wheels were impractical.

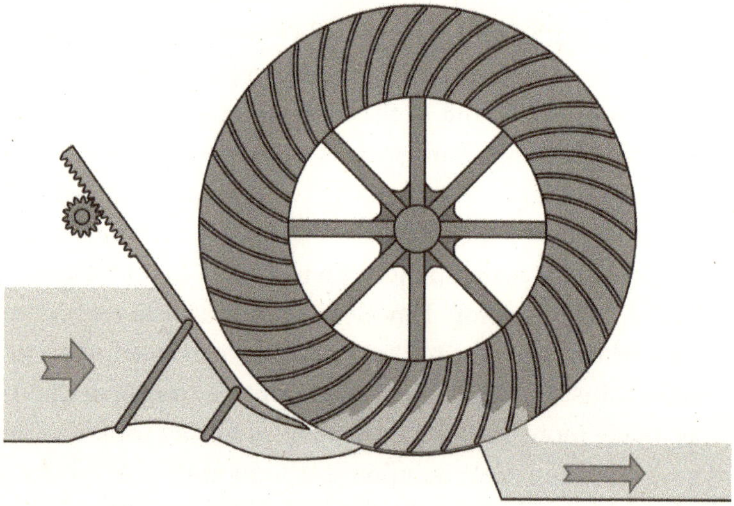

The Poncelet wheel. Water coming in rises smoothly up the curved blades, powering the motion, before falling back down again. Since the wheel moves around at half the speed of the incoming water, by the time the water comes back out it has zero horizontal speed and has given all its energy to the wheel.

The Société d'Encouragement pour l'Industrie Nationale offered a prize of 6,000 francs to any engineer who could develop a practical large-scale vertical axis wheel, and this set off a frenzy of innovation. One of the first entrants was an engineer from the Royal School of Mines, Claude Burdin. He produced a number of designs for machines with curved blades, like Poncelet wheels, and he coined the name "turbine" for them from the Latin *turbo*, meaning a whirlwind, or tornado. In the event, Burdin's turbines were never built, but he inspired one of his students, Benoit Fourneyron, who developed and built a series of outflow turbines. Water was piped into a ring-shaped chamber from above, where it was deflected sideways by a set of fixed curved blades, onto a second set of curved moving blades, which it set spinning before being ejected outward through holes on the sides. He was able to pipe water into these devices to exploit waterheads from as low as

5 feet (1.5 metres) to over 500 feet (150 metres), devices that worked at pressures of over 15 atmospheres. The higher the head, the faster the turbines spun, and the more power they could produce. A motor he designed for a textile mill in the Black Forest was only twelve and a half inches (32 centimetres) in diameter and weighed just 40 pounds (18 kilograms), yet the 360-feet (110-metre) head of water set it spinning at up to 2,300 rpm—almost forty times a second. It developed 60 horsepower at an efficiency of over 80 percent, a third of the power of the Laxey Wheel at a tiny fraction of its size and cost!

Fourneyron was awarded the Société's prize, but there was one problem with his outflow designs: they worked well only at full power because if water flow was reduced, the chambers emptied and the efficiency of the device fell dramatically. So his success stimulated further experiments on turbine design not only in France but also in another territory whose expanding economy demanded new sources of power: the United States. By the middle of the nineteenth century, New England industrialists had already exploited the energy of the lowland rivers, draining the Appalachian Mountains into the Atlantic, using waterwheels to power a range of flour mills, sawmills, and textile mills. Like the French, they sought new methods of extending their industry into the foothills of the mountains; and competition drove rapid developments in technology in the 1830s and 1840s. These culminated in the turbine design that has since proved to be the most influential and important across the world.

Two American engineers, Samuel B. Howd and Uriah A. Boyden, had, in the 1830s, produced a range of turbine designs, from improved Fourneyron outward-flow machines to an inward-flow design by Howd himself; and they passed the data from their experiments on these machines to an Anglo-American engineer, James B. Francis. At the time, Francis was working as the chief engineer at the Proprietors of Locks and Canals for the Lowell Manufacturing Companies, having succeeded his former boss and mentor, the railway engineer George Washington Whistler (father of the painter James Abbott McNeill Whistler). Francis quickly realised that inward-flow turbines were generally more efficient than outward-flow ones, and for obvious reasons.

If you set water to spiral inward within a snail-shaped casing, and turn a series of rotating blades as it does so, it will strike the fastest-moving part of the blades, the tip, first, reducing any shock. It will then be gradually slowed down as it moves inward towards the centre of the device, where the blades are moving more and more slowly. By the time it has reached the centre, the water will have lost almost all its energy, so the efficiency of the turbine should be high. When designing a turbine along these principles, Francis used calculations and experiments to devise the optimal shape for the blades. He included a set of stay vanes to divert water into the device at the correct angle; he curved the moving blades back along their length, to smooth the flow; and he twisted them along their length so that they forced water gradually to the side as it moved inward. This enabled the slow-moving water that reached the centre of the device to run out of it at right angles. Francis turbines were soon reaching efficiencies of over 90 percent and proved ideal to exploit a range of water heads from 30 to 600 feet (10 to 180 metres). By the end of the century they had become the dominant machines to harness the power of water, as they still are today.

Elsewhere in the world, engineers also continued to develop novel hydraulic devices, even in Britain, which had plentiful supplies of coal to fire its steam engines. For instance, in 1850, James Thomson (brother of the more famous scientist William Thomson, who became the Lord Kelvin we met in chapter 4), working in Northern Ireland, one of the wettest parts of the country and with the poorest access to coal, designed his own inward-flow turbine, the Vortex. Thomson also realised that if the spin of an inward-flow turbine was reversed, with its blades rotating backwards, it would act as an efficient pump; its blades would gradually and smoothly speed up water, moving it from the centre to the edge of the device. The idea of the centrifugal pump was born. Such machines could pump water faster and more efficiently than the reciprocating pumps that had been used up to that time and raise it to huge heights. The first commercially successful centrifugal pump was the Appold pump, developed and gifted to the world by the wealthy philanthropist John Appold, and first shown in the Crystal Palace at the Great Exhibition of 1851. Set to work powered by steam

The operating principle of the two most important types of turbines used with high levels of water. In the Francis turbine (*top*), water entering the snail-shaped casing spirals inward, rotating the curved blades before running out at the centre. In the Pelton wheel (*bottom*), a nozzle directs fast-moving water into a set of curved cups that slow it and divert it back the way it came. Since the wheel moves at half the speed of the incoming water, the water has zero net velocity as it comes out.

engines, Appold pumps soon proved ideal for marsh reclamation, draining the last remaining lake in the East Anglian fens, Whittlesea Mere, in 1853.

A modern centrifugal pump. Water is sucked in at the centre and the rotating blades force it around a snail-like whorl and outward into the outlet pie, on the top right. Note that this is the reverse of the action on a Francis turbine.

Though they are highly efficient, Francis turbines had the disadvantage that, working under pressure, they were rather heavy and had to be permanently plumbed in. So the prospectors who were combing the Rocky Mountains of the US West for gold and silver had to develop a quite different form of turbine to power their mining operations. There were plenty of mountain streams in the area, of course, and these were quickly used to help excavate the land; the miners diverted the water from the streams into tubes that stretched down the hills and ended in nozzles, which they used to produce a jet of water to wash away silt and sand from the grains of metal. The miners realised that they could also use these jets of water to power their own little wooden waterwheels.

The "hurdy-gurdy" wheels they devised were simple structures made from triangular blocks of wood 4 inches (10 centimetres) wide sandwiched between wooden wheels. The miners directed the water flowing out of the nozzles, spraying fast-moving jets of water into the sawlike grooves of the wheel, spinning it at high speeds and generating large amounts of energy, as well as a good deal of spray.

Hurdy-gurdy wheels provided the vast majority of the power used by miners in the 1850s California gold rush and the Nevada silver rush of the 1860s, but the miners soon realised that they could improve their efficiency, and at the same time reduce spray, by using curved blades as in Poncelet's waterwheel. If they directed water into one side of a semicircular bucket, it would be slowed down more gradually as it neared the centre and come out at the other side at the same speed. If the wheel was set running at half the speed of the water jet, as the water came out, it would have zero net speed and would have given up all its energy to the device. It would simply drop to the ground and there would be no spray. And if the miners split the water at the centre to go into two such cups set side by side, the lateral forces each cup produced would cancel each other out. Engineers worked hard to make the device practical, the most successful being Lester Pelton, a former gold miner who had moved to Nevada. He made a fortune marketing his Pelton wheel. Pelton's rival, William A. Doble, made further improvements in the shape of the buckets, and after Pelton's death, the two firms combined and started to export Pelton wheels around the world, so that by 1900 more than eleven thousand were in use. Like Francis turbines, they were able to operate at efficiencies of over 90 percent and could produce up to 10,000 horsepower, fifty times the output of the Laxey Wheel. They were the turbine of choice to exploit water drops of more than 300 feet (100 metres).

The new turbines were cheap, light, and highly efficient, but since they worked best with large falls of water, they were only really suitable for hilly, mountainous areas. To assure a permanent energy supply, the water also had to be held behind huge dams that were built across river valleys, which added to the cost and environmental impact of hydraulic power sources. So at first they did not help to provide energy in towns and cities, which tend to be situated on mature rivers or around the coast. A new form of energy was clearly needed. The first novel energy source that was used to power homes was coal gas, which could be produced by heating coal, and was pumped around cities inside a network of pipes. But gas was dangerous and dirty and was mainly used to provide heat and light. Instead it was electricity, probably the first practical innovation to have been developed directly as a consequence of the scientific revolution, that took over as the primary energy source. And like waterpower, electricity depended on new, rapidly rotating machinery.

As we saw in chapter 4, Michael Faraday showed in 1821 that he could convert rotational motion into electricity simply by moving a conducting plate through a magnetic field produced by a horseshoe magnet. His device was extremely inefficient because outside the influence of the magnet, the electricity could move back across the plate, reducing the current and heating the plate, so that it produced a very small voltage. Soon, however, it was realised that this problem could be overcome by separating the metal into coils of insulated wires. The first practical dynamo was demonstrated by the Frenchman Hippolyte Pixii in 1832, and dynamos were developed into DC generators by engineers, many of whose names are still famous: Werner von Siemens, Charles Wheatstone, and S. A. Varley.

As generators were made larger they became more powerful and their efficiency reached values up to 99 percent, so producing enough energy for a town became practical. And once the generators had produced electricity it could be transmitted to wherever it was needed along thin metal wires, a far simpler and cheaper arrangement than the array of axles, gears, and belts that were needed to transmit mechanical power. The first mains electricity was produced by driving generators using steam engines, and cities such as New York were being supplied with DC power from as early as 1880. But steam engines had to be geared up to drive the generators, and their huge size and low power output made power stations unprofitable. Far more powerful and smooth-running machines were clearly needed before electricity could become a practical way of powering whole cities and countries. The answer, once again, was to devise a new type of rapidly rotating power source.

The key was to use steam not to drive pistons up and down cylinders, but to power fast-spinning turbines. Unfortunately, engineers could not simply transfer water technology directly, to make steam-powered equivalents of Francis turbines and Pelton wheels; steam is a gas, and as it does its work it not only slows down but also expands and loses pressure, so using steam in a water turbine would use only a small proportion of its energy and would be highly inefficient. To harness all its energy, steam instead has to be fed in multiple stages across a series of rotors that get larger and larger as the steam expands through the process. And just as the energy in a moving stream of water can be extracted either by firing it through a nozzle or using it under pressure, there are two different ways of exploiting steam power.

The Swedish engineer, Gustav de Laval, devised an impulse turbine that used the same principle as the Pelton wheel. He developed a nozzle that accelerated the steam to supersonic speeds and diverted it into a series of cup-shaped blades on a spinning rotor. The advantage of de Laval's machine was that it did not need to be pressurised, so it was light and cheap, but it spun at incredibly high speeds, up to 30,000 rpm. Consequently, it had to be geared down to run a generator, and lubricating the bearings was tricky, particularly as steam contaminated the oil. It ultimately lost out to devices that used steam under pressure, but de Laval's work did have important side benefits. The shape he developed for his nozzles is replicated in all modern rockets. And de Laval also invented a spinning centrifugal separator to remove water from his oil, a separator that he then went on to customise for the dairy industry. Together with the engineer Oscar Lamm, he produced a hand-powered milk-cream separator, which greatly reduced the time farmers needed to make butter. The principle of separating fluids of different densities by spinning them at high speeds is the basis of all modern centrifuges, most notably being the method used to separate uranium 238 from uranium 235 to produce the fuel for nuclear reactors.

In contrast to de Laval's machine, the steam turbine devised by the British engineer Charles A. Parsons was a reaction turbine that works under high pressures and uses the principle of the propeller. A Parsons turbine consists of an axle that holds several rows of free-spinning fan blades, like those on a Wild West wind pump. The steam moves along the axis of the turbine and hits the blades, causing them to rotate. Of course, this tends to make the steam rotate in the opposite direction around the axle, so before it meets the next set of blades, it is diverted by a series of static blades, or stators, back into the axial direction. As steam travels down the device and loses pressure, it expands, so the blades get larger and larger the farther down the turbine they are, allowing more of the energy to be extracted from the steam. The exhaust gases can also be piped to a larger lower-pressure turbine and sometimes even to a third one to extract the remaining power.

Parsons's ship Turbinia *at speed*. The turbines drove the ship at over 40 miles per hour.

Parsons turbines need to withstand high-steam pressures, so they have to be heavy, but they are much more powerful than a steam engine of the same size. They also proved to be more efficient than de Laval turbines, and since they operate at more manageable speeds of around 3,000 rpm, they were more reliable. At first, Parsons had difficulty interesting the British establishment in the worth of his turbines, not the first or last to have problems with the poor technical education of those in power, so he set about staging a spectacular demonstration of their superiority. He mounted four of his high-powered turbines onto a narrow-hulled ship, the *Turbinia*. After some tinkering with the design of the propellers, which were the first to be spun around so quickly, the vessel became the fastest ship afloat, capable of speeds of 40 miles per hour (64 kilometres per hour). The *Turbinia* turned up, unannounced, at the diamond jubilee naval review in 1897, speeding between the lines of battleships, easily evading interception and humiliating the admiralty. Impressed despite themselves, the navy quickly took to mounting steam turbines in its ships, and merchant vessels and ocean liners were soon afterward routinely fitted with them. They retained their primacy right up until the Second World War.

Parsons turbines also had the merit that, as they got bigger, they became not only more powerful but also more efficient, just like electricity generators. And since they spun at rates of 3,000 to 4,000 rpm, they were ideal to drive a new generation of AC generators, producing electricity at 50 or 60 hertz. Coal- and subsequently oil-fired power stations got larger and larger, their huge generators being driven by gigantic Parsons turbines. Visitors to the Tate Modern art gallery on London's South Bank are often overwhelmed by the 500-feet-long (150-metre-long) turbine hall. This space once held the turbines and generators of what used to be Giles Gilbert Scott's Bankside Power Station, which produced a peak output of 200 megawatts, fifteen hundred times the capacity of the Laxey Wheel. But the Bankside Power Station was itself small compared with later ones. The huge Drax Power Station near my home, which opened in 1974 and was formerly fed by the coalfields of South Yorkshire, is 1,300 feet (about 400 metres) long and holds turbines capable of generating almost 4 gigawatts—thirty thousand times as much as the Laxey Wheel and 7 percent of the total UK demand. And as engineers developed nuclear power in the 1950s, they found that steam turbines were the ideal way to convert the heat produced by nuclear fission reactions into electrical power. The heat is transferred from the nuclear pile into steam using heat exchangers and fed to the turbines. Between coal, oil, biomass, and nuclear power stations, around 55 percent of the world's electricity is today generated using steam turbines.

The ability to generate vast amounts of energy using spinning steam turbines and electrical generators and to transmit it huge distances has had a massive impact that transformed the world in the twentieth century. It certainly went far beyond just lighting and heating our cities. The pioneers of electricity found that they could reverse the action of the electrical generators to produce electric motors, just as hydraulic engineers could reverse the action of turbines to produce pumps. The first electric motor was designed in 1827 by the Hungarian physicist Anyos Jedlik, and the first commercially successful models went into production in the 1860s, coinciding with the launch of generators. They can be made in almost any size and used to power devices that were previously driven by hand. For instance, my mother had a treasured Singer sewing machine from the 1930s that was almost identical to earlier models that had been driven by a treadle. It simply had a small electric motor that drove the machine via a rubber belt. In modern factories, each machine is powered by its own independent motor, so the machines can therefore be arranged in a vast single-story building, forming long production lines—very different from the multistory mills of the nineteenth century with their dangerous and cumbersome arrangements of shafts, cogs, and belts. Nowadays, around 50 percent of the electricity we generate is used to power electric motors, including a new generation of electric vehicles, as we shall see in the next chapter.

The huge and growing demand for electricity in the early years of the twentieth century called for more ways in which it could be produced, so engineers returned to the sources that they had previously exploited directly: water and wind power. They were helped in this enterprise by the development of AC electricity at the end of the nineteenth century. Since AC current can readily be converted into high voltages using transformers, and transported long distances with little energy loss, it became possible to generate power even in isolated districts for use in major towns.

The first hydroelectric power had in fact been produced as early as 1873 by the British engineer and armaments manufacturer William Armstrong at his country house, Cragside, in Northumberland. Powering some of the first electric lights, the electricity was generated using an Archimedes' screw that exploited the fall of water behind a dam in his landscaped gardens. Recently restored, it can be seen working in the beautiful valley under the shade of huge Douglas firs, nature's own hydraulic engineers. Soon, engineers started to exploit the power available in the mountainous areas of Europe, especially in Scandinavia and the Alps, and in the mountains of the United States, using Francis turbines and Pelton wheels to drive their generators. An early example can be seen at the Lake Vyrnwy reservoir in North Wales, which was built in the 1880s to supply the city of Liverpool with drinking water. The power installation could be regarded as the swan song of Victorian whimsy, as the two 6-feet-wide (2-metre-wide) Francis turbines, which together developed some 120 kilowatts (roughly the same as the huge Laxey Wheel), were housed in what looks like a tiny stone cottage.

In the early years of hydroelectricity, much of it was used locally to provide cheap electricity to smelt aluminum, which is why this industry is still concentrated in the uplands. Since then, however, countries around the word have exploited their water resources to produce huge amounts of hydroelectricity that they transmit to their major cities. Modern hydroelectric schemes usually involve the construction of ever larger dams to hold back and control rivers for industry, irrigation, and drinking water, and they generate electricity using high-power hydraulic turbines. The American West is largely powered by gigantic dams; the Hoover Dam, built on the Colorado River during the Great Depression, can produce over 2 gigawatts with its Francis turbines, while the Grand Coulee on the Columbia River, completed in 1974, produces almost 7 gigawatts. The largest scheme of all is China's Three Gorges Dam on the Yangtze River, whose Francis turbines are capable of generating an astonishing 13 gigawatts of electricity. Hydraulic schemes continue to be built around the world, despite the environmental problems that they bring, and today they account for around 16 percent of total global electricity generation.

Though they are brilliant at extracting energy from deep lakes in upland areas, Francis turbines are inefficient at extracting energy from lowland rivers and estuaries. In the years before the First World War, therefore, the Austrian engineer Viktor Kaplan investigated alternative ways in which he could convert the kinetic energy of this low-pressure water into mechanical power and electricity. He realised that if he made the water flow through tubes he could remove most of its energy by intercepting it using a rapidly spinning propeller. And if he used variable-pitch propellers, he could make them work efficiently whatever the speed of the water. From the 1920s onward, Kaplan turbines were quickly developed to obtain energy from falls of water below 30 feet (about 10 metres). The French have been particularly enthusiastic adopters of these turbines, using them in barrages and weirs across many of their major rivers; the La Grande-1 power station in Quebec, Canada, can produce almost 1.5 gigawatts. Another large installation using Kaplan turbines is the Rance tidal power plant in Brittany, built where the tides are amplified by the constriction of the English Channel, and diverted by Coriolis forces south into the mouth of the Rance estuary. Opened in 1966, it can produce a peak electricity output of 250 megawatts. Finally, to exploit intermediate water drops of 60 to 160 feet (20 to 50 metres), the engineer Paul Deriaz produced a turbine with propeller-like blades, though in his design they were swept back at 45 degrees. Deriaz turbines both look and operate in a way that is intermediate between a Kaplan and a Francis turbine, and they are most notably used at the Sir Adam Beck generating station at Niagara Falls, which can produce a peak output of almost 2 gigawatts.

Engineers are now seeking to extend the use of water turbines into the open sea. The tidal flows around our oceans are a vast potential source of energy, particularly when they are concentrated between islands. The largest free-floating tidal-powered turbine built to date is the experimental Orbital O2, a 240-feet-long (73-metre-long) submarine-like structure that is tethered in the Fall of Warness in

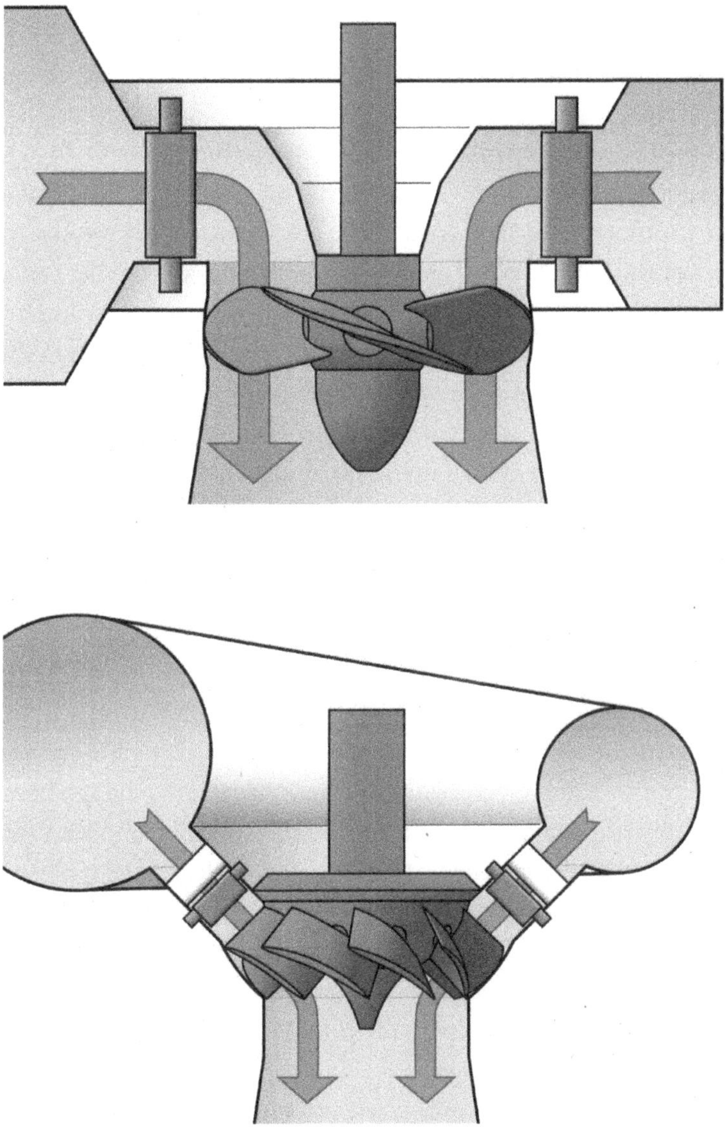

Operation of low-pressure turbines. The Kaplan turbine (*top*) consists of a variable-pitch propeller inside a tube that intercepts water flowing axially. The Deriaz turbine (*bottom*) acts halfway between a Kaplan and a Francis turbine; its blades intercept water flowing at 45 degrees inward.

the Orkneys off Northern Scotland. On its submerged nacelles it holds two 65-feet-diameter (20-metre-diameter) rotors that can produce 2 megawatts of electricity. The power output is tiny, but it is hoped that this device can act as the prototype for larger, more powerful units in the future.

In the last thirty years, engineers have also finally begun to exploit the massive power of the wind, not only on the tops of hills, but also in the uninterrupted flat vistas of the sea where winds are strongest and most consistent. Wind turbines resemble huge propellers and act like Kaplan turbines, but since the wind cannot be diverted into tubes, wind turbines cannot extract all the energy that passes through them; the maximum theoretical efficiency is only 59 percent. Nevertheless, wind turbines are so large that wind farms consisting of hundreds of these devices are capable of extracting huge amounts of energy from the wind. The largest wind farms in the world are currently being constructed in the shallow North Sea off the coast of Yorkshire where I live. The turbines themselves are enormous. With blades 250 to 280 feet (75 to 85 metres) long, they can tower almost 650 feet (200 metres) above the waves, twice the height of the Statue of Liberty. A single blade, set up for Hull's City of Culture celebrations in 2017, spanned the width of its largest public square, dwarfing the surrounding buildings. Each turbine can produce 7 megawatts, though because they spin so slowly they need a gearbox to speed up the rotation so that they can power the generators in their nacelles. Renewable power from wind turbines is now a major player in world-wide electricity generation; output rose to 740 gigawatts in 2021, around 5 percent of global electricity production, and is increasing at a rate of around 10 percent a year. By 2050 it could provide 6,000 gigawatts of electricity, 35 percent of the world's needs.

Of course, there is a problem with electricity generated by renewables, as river flow and particularly the tides and winds are intermittent. If we are to rely on them for our future energy needs we will need a way of storing energy. Batteries are one solution, but they are expensive to construct and demand huge amounts of precious metals such as lithium. Instead, we could use other methods that once again

Turbines used to harvest the energy of the wind use huge propeller blades.

use turbine technology. As we have seen, Francis turbines can also be used as centrifugal pumps and electricity generators can be used as motors. Engineers have learned to make use of this fact. On one of our family holidays to the mountains of Snowdonia in the 1970s, we visited one of the largest of the power stations that make use of these capabilities. Dinorwig sits on the site of a huge former slate mine and uses two lakes that lie on the slopes of Elidir Fawr, one at a height of around 2,000 feet (610 metres), the other around 1,600 feet (500 metres) lower. At peak times, water is allowed to flow down through pipes from the upper reservoir into three Francis turbines, producing almost 2 gigawatts of power, and released into the lower reservoir.

The flow is then reversed at night when demand is low by using the turbines as pumps. The total efficiency of the operation is only around 75 percent, so overall the operation actually uses energy. However, it acts to smooth energy demand so that only the most efficient power stations ever need to be used.

The potential to expand pumped-storage schemes such as Dinorwig is limited due to the lack of suitable sites, so in recent years researchers have investigated other potential methods to store energy. One promising technique is compressed-air energy storage (CAES), which, during periods of high-energy supply, use electrically powered turbo-compressors to compress air, and at times of high demand allow it to expand again, using turbo-expanders to obtain energy from it. The turbo-compressors and turbo-expanders are both rotating devices and can be fitted with propellers, like Kaplan turbines, or rotary blades, like Francis turbines. Currently, the efficiency of CAES is low, because of the temperature changes that occur when gases are compressed or allowed to expand. The number of sites where it can be used are also limited; they are usually placed above abandoned salt mines, which can hold enough compressed air to store reasonable amounts of energy. For this reason, another technology, liquid air energy storage (LAES) is currently being investigated. If air is cooled enough to liquefy it, its volume is greatly reduced and it does not need to be held under pressure. It can be stored in huge purpose-built silos. LAES plants could therefore be sited almost anywhere. Like CAES plants, they currently suffer from low efficiency, around 25 percent, but this can be improved to 70 percent if they are colocated with a low-energy cold store, such as a gravel bed, and a source of waste heat, such as a landfill or conventional power station. An experimental 50-megawatt CAES plant is currently being built in Greater Manchester, with another plant planned for northern Vermont, in the United States.

Today, therefore, we are dependent for our electricity on fast-spinning devices, and this is likely to be the case long into the future. Almost all of our electricity is produced by rotating generators, and most of these are powered by turbines of some sort, driven by steam, water, wind, or as we shall see in the next chapter, gas. Only solar cells produce electricity without a generator, and currently that accounts for a mere 2 percent of global supply. And of the electricity we produce, over half is used to power rotating motors, many of which in turn drive rotating devices. Pumps, for instance, account for 10 percent of the world's energy consumption. Spin powers our whole world.

CHAPTER 12

Going for a Spin

As anyone who is familiar with novels written at the turn of the twentieth century will know, travel at the time was a curious hybrid affair. Henry James's wealthy heroine, Isabel Archer, could travel between the United States and Europe on steamships and cross the continent on intercity express trains. Lowly clerks such as Mr. Pooter in George and Weedon Grossmith's *The Diary of a Nobody* could travel into the city of London on commuter trains. But railway tracks block roads, limiting their use in cities, and because of the low friction between iron wheels and rails, trains with conventional wheels could only travel on flat routes through the rural lowlands; railways had to be engineered across valleys and through hills using expensive bridges, cuttings, and tunnels. Even in the lowlands, where they were easiest to build, railways were therefore expensive to construct and were consequently few and far between. So when Isabel Archer visited her many unsuitable admirers at their country estates, or Sherlock Holmes visited his clients at their country houses, they had to be picked up at the nearest station by a pony trap and driven the last few miles; the locals went around on horseback, on the backs of carts, or on foot. Back in London, Archer was driven in carriages, while Holmes travelled to and from Baker Street or chased Moriarty around the city in horse-drawn hansom cabs. This all demanded an infrastructure for the horses. Aristocrats had stables behind their grand town houses, stables that have in more recent years been converted into posh mews. Tradesmen also moved their goods around on carts and wagons, and there were a few horse-drawn street omnibuses pulled by exhausted nags. In the 1890s, therefore, London had over three hundred thousand horses for transport alone, and New York around five hundred thousand. And with horses came inevitable problems. In 1894, the *Times* of London cried out, "In 50 years, every street in London will be buried under nine feet of manure." This became known as the Great Horse Manure Crisis. The world's first urban-planning conference, which followed soon after in 1898, was unable to come up with a solution.

Yet within twenty years the problem had been solved. Developments in the design of wheels, chassis, engines, and roads had enabled engineers to build human- and motor-powered vehicles that could travel faster and more freely along the existing road network, while propellers had enabled airplanes to break free from the earth's surface. Within seventy years, jet planes could fly faster than sound, and rockets enabled astronauts to leave earth's orbit. My grandfather, brought up in Sherlock Holmes's London, lived to see the emergence of motorways and intercontinental jet travel, and watched the first men land on the moon. All of these achievements were founded on revolutions in spinning technology.

The forerunner of all these vehicles was the ultimate in personal transport: the bicycle. They make a notable appearance in the Sherlock Holmes's 1903 short story "The Adventure of the Solitary Cyclist," but it had taken eighty years for bicycle technology to advance to the stage when bicycles were common enough to start to influence popular culture. The forerunner of the bicycle, the velocipede, or hobbyhorse, appeared as early as 1818. It featured two wheels, a saddle, and front-wheel-steering with handlebars, like a modern bicycle, but had a wooden frame and wooden wheels like a carriage, and the rider propelled it by running their feet on the ground, like an astride version of a child's scooter. There was a brief craze for these toys among fashionable young men, as there has been recently for Segways and electric scooters; like them, velocipedes were involved in a series of accidents, which helped put a quick end to the fever. The first person to affix a crank and pedals to the front wheel of a bicycle was the German Philipp Moritz Fischer in 1853, and by the 1860s commercial bicycles were being made in France, where they were known as velocipedes, and in the United States and Britain, where they were known as bone-shakers because of the shocks transmitted by the rigid wheels and iron tyres. But these machines were heavy and had a limited speed because of the small size of the wheels; to travel quickly the rider had to pedal ridiculously fast. Gradually these problems were overcome. In 1869, the Frenchman Eugène Meyer introduced the wire-spoke tension wheel, capitalising on an 1832 invention by the British aviation pioneer Sir George Cayley, and along with the Briton James Starley, developed tubular-framed bicycles with huge front wheels: penny-farthings, or more prosaically, ordinaries. Used by sporting young men for racing, these machines were fast, but extremely dangerous. The rider was positioned high up above the ground, which was hazardous in itself, but the main problem was longitudinal stability. If the rider braked quickly, or pushed down hard on the pedals, the inertial or propulsive forces acted above and in front of the point at which the wheel touched the ground, and he was all too apt to "come a cropper" or "take a header," falling forward over the front wheel. Some penny-farthings were made with a complex crank arrangement to allow the pedals to be positioned farther back, but these were never popular and bicycle evolution turned in another direction.

From the 1830s onward, inventors sought to develop ways to drive bicycles through the rear wheel. The Scottish blacksmith Kirkpatrick Macmillan is said to have been the first person to build such a bicycle, by adding cranks to the front forks, which were joined to the spokes of the back wheel. By swinging their feet back and forth, riders could propel the machine. However, this arrangement was awkward, and rear-wheel bicycles did not become common until the invention of the safety bicycle, which was developed in the 1870s and 1880s. In such machines, the rear wheel was driven by a chain powered by cranks mounted at the bottom of the frame, as in modern bicycles. The chain ran over a larger sprocket on the front than on the rear, gearing up the speed of rotation and removing the need for a large rear wheel. The rear wheels were also fitted with a freewheel, a gear with stepped teeth that allowed the wheel to turn forward even if the rider had stopped pedaling. Shod with solid rubber, or from the 1890s, with pneumatic tyres, which greatly smoothed the ride, safety bicycles took off in popularity, with both men and women.

In the late nineteenth century, inventors were continually coming up with new devices to improve bicycles. In 1869, for instance, the French mechanic Jules Suriray patented the use of ball bearings in bicycles. Ball bearings reduce friction because rather than sliding past the axle, the housing of the wheel rolls along on a ring of steel balls, like the supposed sleds of the ancients. The main downside of the arrangement is that like the tree trunks below a moving stone, the balls are liable to jam up against one another, so they have to be caged, with the holes in the cages acting like tiny axles for the balls. Despite the friction this produces, ball bearings have since revolutionised machinery, allowing the development of the rapidly spinning turbines and generators we examined in the last chapter. And they soon showed their worth in bicycles, as they were mounted on the wheels of the penny-farthing ridden to victory by James Moore in the first-ever bicycle road race, the Paris-Rouen, in November 1869. Other improvements soon followed: gears to enable cyclists to climb hills, brakes, and more comfortable saddles. Reading Jerome K. Jerome's *Three Men on the Bummel*, a comic travelogue about a cycling holiday in Germany written in 1900, it is plain that, like the MAMILs (middle-aged men in Lycra) of today, late Victorians were obsessed with cycling gear. But it was not only young men who enjoyed cycling. The safety bicycle enabled women for the first time to travel without chaperones; transformed their clothes, forcing them to wear bloomers rather than skirts; and enabled longer-distance and even interclass courtships, as H. G. Wells's comic novel *The Wheels of Chance* attests.

Once again, however, it is worth pointing out that none of the advances that made cycling possible were made by scientists: it was all practical men. And even today, scientists seem confused about why bicycles are stable. Many physicists have cast their minds to tackle this problem mathematically, usually without providing much in the way of illumination. One common idea is that bicycles are stabilised by the gyroscopic action of the front wheel. If you roll a hoop along the ground, like a Dickensian street urchin, as it starts to lean over, precession will tend to turn it inward towards the lean, preventing it from falling over; it will roll in an inward spiral path, only finally toppling over when it stops moving. The problem with this idea is that bicycle wheels are extremely light compared with the rider, so the effect is very small. The real reason for the stability of bicycles was demonstrated in 1970 by an industrial chemist from Britain, David Jones. Rather than just do the mathematics, he performed a series of brilliant and increasingly hilarious experiments. In an attempt to produce the "unrideable bicycle," he added contrarotating wheels to the front wheel to eliminate the gyroscopic effect, experimented with tiny caster wheels, reversed the front forks, and even extended them. He finally showed that the main stabilising force is due to the fact that the front wheel touches the ground behind the line of the front forks. Not only does this supply stability because the ground reaction force on the rim acts behind the steering, like in a trailer, but leaning the bicycle causes the wheel to automatically turn inward, as this lowers the bicycle's centre of gravity. Jones's only disappointment was when he modified a bicycle so that its front wheel projected ridiculously far forward, eliminating the stabilising effect, it was not quite as unrideable as he had hoped. But perhaps this was fortunate, as it enabled him to survive to write up his findings in what must be one of the most enjoyable-ever physics articles.

Despite their dynamic stability, bicycles are still tricky to ride, and being limited in the loads they can carry, mass bicycle transport has only really taken off in the flatlands of Holland and Denmark, and in Maoist China. Elsewhere, engineers started to develop powered vehicles with several wheels. Many cities looked to convert their horse-drawn omnibuses and trams to motorised traction. American cities tended to opt for cable cars, hitching their vehicles to continuously moving underground cables that ran beneath the tracks, and which were directed around the streets using a complex network of pulley wheels. Cable car systems had the great advantage that they could operate even in the hilliest districts, but the high friction in the pulleys wasted over 90 percent of the energy supplied by the stationary steam engines that powered them. Consequently, by the end of the nineteenth century, most of these systems had shut down, and only a few lines remain in use, for instance in San Francisco, where they are essentially tourist attractions. In Europe, most cities opted to convert horse-drawn trams first to steam operation, and then most successfully to electricity. The electric motors were supplied with mains electricity using a pantograph on top of the tram which slid along an overhead power line, just as in most modern electric trains. Trams, or streetcars as they are called in North America, remained popular until after the Second World War, and have had a renaissance in recent years, but the early vehicles were jerky and uncomfortable, and they were still tied to their limited track network. The future of land transport, both in cities and the countryside, proved to be motor vehicles.

The inventors of the first cars and vans had the advantage that they could exploit the advances that had been made by centuries of carriage builders and decades of bicycle designers. They could make the frames of their cars using lightweight rolled steel tubing, mount the frame on leaf springs, and use wire-spoked wheels with pneumatic tyres. The one remaining difficulty was how to power the rear wheels so that as the car went around a corner the outer wheel could travel faster than the inner wheel. Fortunately, once again, this problem had already been solved, in the form of the differential gear. In this arrangement, the rear wheels are powered by a driveshaft that ends with a rotating cog. But the cogs that drive the left and right wheels are not attached to the driving cog directly, but to two intervening cogs, so that if one wheel encounters greater resistance, the other wheel rotates faster. This form of gear may have been invented as early as classical times; it appears to have acted as part of the arrangement of cog wheels in the Antikythera mechanism, which seems to have been some sort of early calculator. It may also have been used by the early Chinese dynasties as an integral part of their south-pointing chariots. Lacking compasses, the Chinese used two-wheeled chariots that were so designed that, whichever way they turned, a central pole turned in the opposite direction, maintaining a fixed orientation towards the south. The differential was finally patented in 1869 by the cycle builder James Starley for use in tricycles, which could be ridden by unadventurous men and sartorially handicapped women. It was Starley's design that was used by Karl Benz in the world's first car, the 1885 Motorwagen, which was essentially just a motorised tricycle. Four-wheeled cars soon followed, with their front wheels steered using the Ackermann steering system, which we examined in chapter 7. And when front-wheel-drive vehicles emerged, they used yet another preindustrial invention, the universal joint, to allow the driven wheel to be turned while it was being powered. This system, in which the two drive rods are joined by two hinge joints arranged at right angles to each other, was invented as early as the mid-seventeenth century. Robert Hooke pictured a universal joint in his *Micrographia* of 1665, and later developed a system with two universal joints in a row, which ensured that the wheel always spun at the same speed as the driveshaft, whatever the angle of steering.

In the early years of motor vehicles, engineers tried many types of engines. Among the most popular were steam engines, which needed to be fired up before travel and have their fire continually stoked, which is why professional drivers have ever since been known as chauffeurs, which is French for stokers. Electric cars were also popular at first, but the weight of the heavy batteries and their limited range were disadvantages that are still proving difficult to overcome. It was the internal combustion engines that ultimately triumphed, in the form of gasoline and diesel engines. Both of these are reciprocating engines like steam engines, but the pistons are driven by explosively burning fuel with compressed air inside the cylinder, rather than by burning it inside a boiler and releasing it into the cylinders. The advantage is that the explosion within the cylinder produces higher pressures and power output than an input of steam, and the engine can be run at higher frequencies. Internal combustion engines can therefore be built with several small cylinders that move in

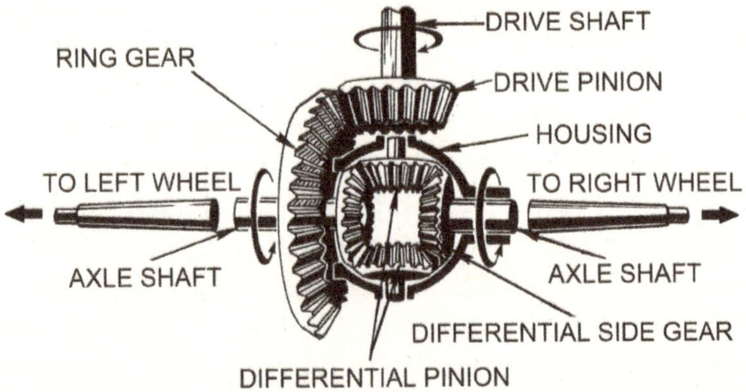

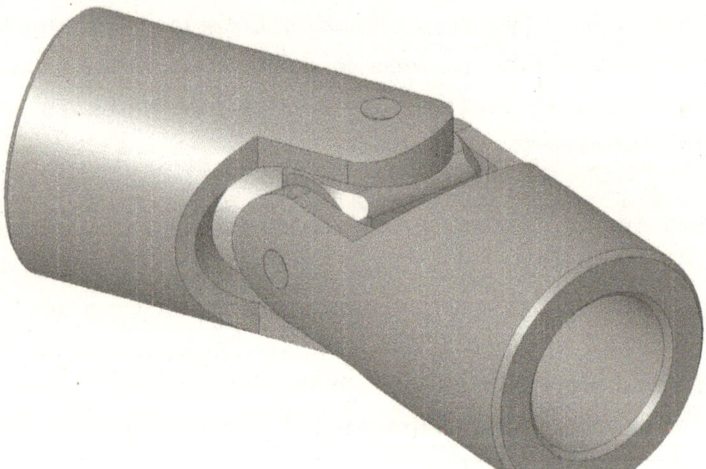

Two mechanisms that have proved crucial for the success of modern vehicles. In the differential gear (top), the axles on the two sides can move at different speeds, allowing a vehicle to corner. The universal joint (bottom) changes the direction of an axle, and if two are used in tandem, it keeps the two ends of the axle spinning at exactly the same speed.

opposite directions at the same time, reducing vibrations, so that internal combustion engines can be far lighter than equivalent steam engines. Powered by these engines and using a wide range of bicycle technology, motorcars soon became popular, especially in the United States, where Henry Ford started to produce the world's first affordable car, the Model T, as early as 1908. The popularity was further helped by improvements in roads. Their smoothness and durability were transformed around the turn of the twentieth century by gluing the stones together with bitumen or tar to produce a smooth asphalt surface.

However, though innovative design had enabled motor vehicles to conquer the roads and skies by the early years of the twentieth century, the same cannot be said of the open country. Arable land continued to be ploughed and harvested using horse-drawn tools, and the numbers of draft horses actually continued to increase, both in Europe and the United States, reaching a peak in the 1930s. And in the First World War, trucks all too easily got bogged down in rutted roads and were hopeless in the shell-pocked hell of the battlefields themselves. Consequently, guns continued to be drawn by horses, and fresh troops had to march miles from the rail terminals of France up to the Western Front. It was only late in the war that French and British military engineers came up with the solution to drive motor vehicles over rough terrain. The caterpillar track was yet another extension of bicycle technology: a huge chain that could spread the load of the vehicle over a wide area and even enable it to cross trenches. The tank was developed independently in the two countries; the British version (the name *tank* derives from the code name for this secret weapon, a "water tank for Macedonia") had huge quadrilateral tracks that ran around the whole vehicle and had cannons mounted in pods at each side. The French version, in contrast, resembled more modern tanks with lower-level tracks and a fully rotating gun turret on top.

In agriculture, the problem of driving vehicles over soft ground was quickly overcome, simply by using bigger, wider wheels, with giant metal or pneumatic rubber tyres. The remaining difficulty was how to use tractors to pull agricultural machinery. The obvious answer was simply to attach a plough or harrow to a tow bar at the back of the tractor, just as they were yoked to teams of heavy horses. The downside to this solution, however, was that if a plough hit a stone or dug too deep into the ground, its resistance to being pulled forward would shoot up. Using a horse team, this was not a problem; the horses would simply stop and wait until the ploughman rectified the situation. Using a tractor, the engine would continue to rotate the wheels relative to the tractor, which would turn over backwards, falling on and sometimes killing the farmer. This was a particular problem for the most popular early tractor, Henry Ford's small Fordson. It was only overcome by yet another former bicycle mechanic, the Irishman Harry Ferguson, who in his youth had built and flown the first successful Irish airplane. In his three-point hitch, the tools were attached via a three-point hydraulic linkage to the back of the tractor. This arrangement meant that a wide range of tools could be rigidly attached, and be raised to take them to the field, before being lowered into the ground for use. The arrangement also moved the centre of gravity of the unit farther back so that the large rear wheels could apply more driving force, and meant that the vehicle could no longer overturn, greatly reducing accidents. Ferguson even introduced a mechanism, draft control, onto the upper mobile attachment point so that if the plough encountered greater resistance the hydraulic system would automatically raise the blades. Ferguson's system revolutionised farmwork so that by the 1940s tractors that used it, including the famous Ford-Ferguson, quickly took over from heavy horses. Motor vehicles with powerful internal combustion engines that drove rubber-shod wheels or caterpillar tracks had at last replaced horses on both road and soil, and these noble beasts were relegated to a largely recreational role.

CHAPTER 13

Taking to the Skies

Even before cars had started to replace horses on the streets of cities, bicycle technology had already helped humans take flight for the first time in heavier-than-air vehicles. The Wright brothers, whose Flyer first took to the skies in 1903, had run a bicycle shop in Dayton, Ohio, and were well-grounded in making lightweight structures. Not only did they carry out wide-ranging tests on a series of gliders, investigating how aircraft could be controlled, but they also devised the optimal ways of driving their aircraft through the skies. They designed and made the first efficient propellers, twisting the blade along its length so that it struck the air at the correct angle; their propellers had an efficiency of almost 70 percent, not far from the 90 percent of the best modern versions. And they drove the two pusher propellers on the Flyer using chain drives from a lightweight internal combustion engine that they had designed and built themselves. Like cars, airplanes were quickly developed into a useful form of transport, their development being accelerated by a series of races, and by the First World War, in which they were turned into formidable killing machines.

Progress in aviation during the first half of the twentieth century was driven largely by the dramatic increases in power that engineers were able to coax from internal combustion engines; output rose from

100 to 200 horsepower during the First World War, to 1,000 to 2,000 horsepower in the Second. To deal with this extra power, aircraft were built stronger and heavier, wood and fabric were replaced by metal construction, and drag was reduced by having a retractable undercarriage and replacing the two wings of biplanes with the single wings of monoplanes. This helped raise the maximum speed of aircraft from around 100 miles per hour to over 400 miles per hour (160 kilometres per hour to 640 kilometres per hour). But one aspect of aircraft performance failed to improve: the ability to fly at high altitudes. The problem was that as air pressure fell higher in the atmosphere, less air would rush into the cylinders at each stroke and their power output fell dramatically. Piston-engine airplanes flew very poorly at altitudes over twenty thousand feet (six thousand metres). Engineers responded by adding pumps to drive air into the cylinders. Some were fitted with superchargers—essentially centrifugal pumps—that were driven around by the engine itself. Others had turbochargers, where the centrifugal pumps were driven by inward-flow gas turbines that were powered by exhaust gases from the engine. This made use of energy that would otherwise have been wasted, making the engines more efficient. Superchargers and turbochargers enabled aircraft in the second half of the war to reach altitudes of over forty thousand feet (twelve thousand metres). In more recent years, turbochargers have also been used to improve the performance of automobile engines. Developments in the 1970s and 1980s in Formula 1 racing cars showed the superiority of turbocharged engines over conventionally aspirated ones, and today many road cars are turbos. Their small piston engines are fitted with two almost identical rotating devices that share the same axle: an inward-flow gas turbine that is driven by exhaust gases and a centrifugal pump that drives air into the cylinders.

But fifteen years before the Second World War, the British engineer Frank Whittle had foreseen that aircraft powered by piston engines and powered by propellers would soon reach the limits of their performance. Not only would they encounter problems at high altitude, but the rapidly spinning propeller tips would encounter sonic shocks well before the airplane reached the speed of sound. Whittle advocated a simple solution to improve the speed and altitude performance of aircraft: to apply turbine technology to the internal combustion engine, just as Parsons had applied it to the steam engine. He realised that if he could compress air using a rotating pump, he could then mix it with fuel in a combustion chamber, to produce a constant stream of exhaust gases that could not only propel the plane at high speeds but also drive the compressor itself around. Since it had just a single rotating part, a gas turbine would be far simpler than an internal combustion engine, and could be made cheaper, lighter, and more powerful.

Whittle decided to compress the air with a centrifugal compressor, essentially an open-fronted centrifugal pump, and to reverse the flow of the compressed air twice before mixing it with the fuel and burning it, to prevent the exhaust gases from moving forward. The British authorities realised that this combination would limit the ultimate performance of Whittle's engines, because the wide centrifugal compressor would have high drag. However, they pressed forward with what was a relatively well-known and reliable technology, and the first flight of a British jet occurred in 1941. Britain's first jet fighter, the Gloster Meteor, became operational in July 1944 and was an immediate success, flying faster than any piston-engine fighter and being able to intercept even the V-1 flying bomb. In Germany, Hans von Ohain had come across Whittle's 1930 patent and together with Ernst Heinkel built a jet plane as early as August 1939. However, Germany decided to develop an alternative technology, using a more complex axial compressor—essentially a line of fans like in a Parsons steam turbine. Consequently, they lost most of their engineering lead, and their first jet fighter, the Messerschmitt Me 262, entered service just a month before the Meteor. Though it could fly even faster, it was not produced in great enough numbers to shift the balance of airpower over Germany.

Ultimately, all subsequent jet aircraft have used axial compressors, like the Me 262, which enabled narrower, more powerful engines to be developed. However, though they are fast, jet-powered aircraft are less efficient than propeller planes because they accelerate the exhaust gases to such high speeds, wasting energy. Consequently, after the

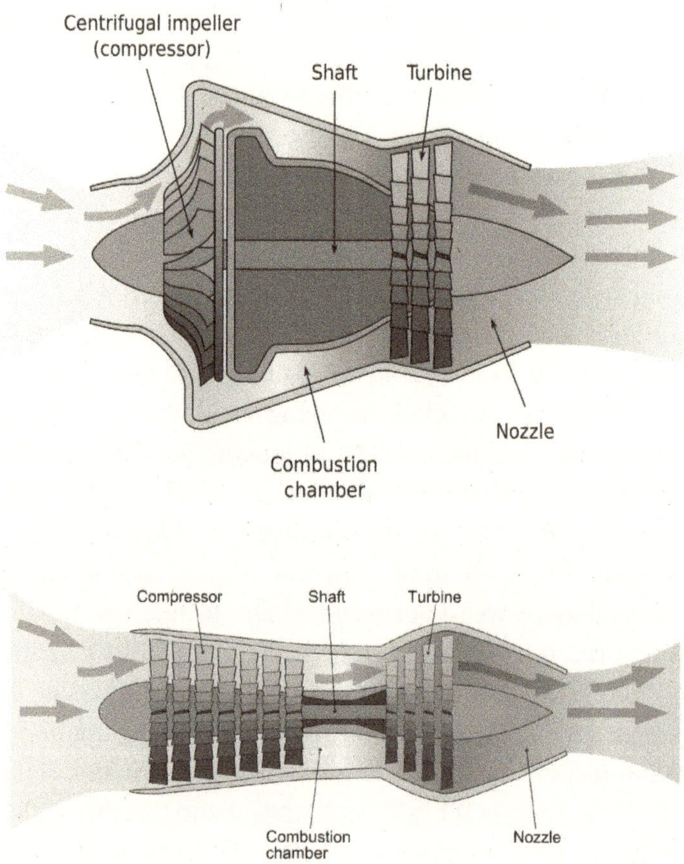

A comparison of the design of jet engines with a centrifugal compressor (top) *and an axial compressor* (bottom). All modern jets use axial compressors, but centrifugal compressors are widely used in helicopters and stationary gas turbines.

war, low-speed aircraft started to be built with turboprop engines, in which most of the energy of the exhaust gases was used to drive conventional propellers. And modern commercial jets use turbofan engines, in which most of the thrust is provided by the huge fan at the front of the engine; only a small fraction of the air it propels backwards is compressed and forced into the engine itself. Because turbofans drive a greater volume of air backwards more slowly than a turbojet, they use far less energy and the cowling around the fan prevents turbulence at the tips, making turbofans far quieter than turboprops.

Gas turbines have proved to be highly flexible and have been turned to many uses. For instance, they are the key power source for aircraft that are capable of taking off and landing vertically. The Hawker Siddeley Harrier "jump jet," for instance, used its Pegasus gas turbine to produce jets of gas that could be directed not only backwards, for forward flight, but also downward, to enable it to take off and land from small ships. The vectored thrust could even be used in combat, to give them unrivaled manoeuvrability. The Pegasus engine was also designed to overcome what would have been a major defect of the aircraft: precession. A rapidly spinning jet engine would normally act as a huge gyroscope, causing the plane to pitch uncontrollably if the pilot attempted a turn when flying at low speeds. The engine was designed so that the blades of the turbine at the rear spun in the opposite direction of those at the front, eliminating any net angular momentum. The successor to the Harrier, the Lockheed Martin X-35B, which has STOVL powers, has similar vectored thrusters at the rear of the plane, but provides the lift at the front when it wants to land using an auxiliary vertical fan that is powered by the main turbine and that rotates in the opposite direction.

Gas turbines are also the preferred power source for the most common vertical takeoff aircraft: helicopters. Lighter than the equivalent piston engine, the turbine of a helicopter typically has a multistage compressor with several axial blades, and a final centrifugal compressor. The energy from the exhaust gases is used to drive the large main rotor. Because their rotors are so large, helicopters have even greater problems than jump jets with precession. One way to overcome this is to use two contrarotating blades, as in the huge troop-carrying Chinook helicopters. Normal single-bladed helicopters not only have to have a small tail rotor to prevent the fuselage from spinning in the opposite direction from the blades, but also have special modifications to their controls. A helicopter is manoeuvred by tilting the axis of the rotor. However, because of the high angular momentum of its large blades, a torque that tried to tilt them forward would cause the rotor to precess like a gyroscope; the blades would actually tilt at 90 degrees from that direction, to the right or left, depending on the direction in which the blades were spinning. Similarly, a torque to the right would actually tilt the helicopter's rotor forward or backwards. The controls of a helicopter are therefore set so that the joystick acts at 90 degrees from the desired direction.

A final example of how gas turbines have been used in transport is in extremely low-flying machines: hovercraft. The concept of a vehicle that could operate over all terrains by being supported on a cushion of air was suggested many times during the first half of the twentieth century. However, the first person to develop a method to put the idea into operation was the British engineer Christopher Cockerell. His solution was to surround the air cushion with a tyre-like skirt that was itself supported by pressurised air, a technique that greatly lowered the power needed to support the vehicle. The largest-ever commercial hovercraft, the Saunders-Roe SRN4 was powered by four Rolls-Royce turboshaft engines, each of which drove a centrifugal lift fan to support the vehicle, and a 90-foot-diameter steerable propeller to drive it forward. The craft held up to 250 passengers and 30 cars. I remember travelling to and from France in one with our family in the 1970s. It "flew" across the English Channel at twice the speed of a conventional ferry and was far quicker to dock, as it simply ran up a ramp onto the harbour.

Gas turbines are also repurposed for a wide variety of stationary uses. Much of the natural gas taken from the world's gas fields is now used to produce electricity. It is burned in huge gas-fired power stations, which produce some 22 percent of the world's electricity output. They are not particularly efficient, because the exhaust gases are so hot, but their performance can be improved by taking off this waste heat and using it to power a steam turbine, in a combined cycle configuration. This can raise overall efficiency to up to 65 percent. The big advantage of gas-fired power stations, though, is that, like jet engines, they can be started and shut down very rapidly, over a matter of seconds. They are therefore vital components of national electricity grids, producing large amounts of the electricity needed during power surges, such as at halftime during televised games of football or during heat waves that require peak air-conditioning.

There is one final use of spinning machinery that has enabled humankind not only to travel rapidly through the air but also to escape the atmosphere entirely and travel to other celestial bodies: in rockets. You would not expect that a rocket would need any such machinery, since a rocket motor itself has no moving parts at all; two propellants simply meet in the combustion chamber, react with each other, and the exhaust gases that this produces are expelled from the nozzle at the back. However, all rockets need some way to pump the fuel into the reaction chamber. Modern rockets are derived in design from the ones used to power the Messerschmitt Me 163 rocket plane and the V-2 rockets produced by the Germans late in the Second World War. As the liquid fuel expands, some of its energy is used to drive a turbine, which in turn is used to drive a pump to move fuel into the combustion chamber. A rocket's turbopump therefore resembles a car's turbocharger. After the war, Nazi rocket engineers such as Wernher von Braun were enticed to the United States, while others went to the Soviet Union; both were set to work in a race to produce the most effective nuclear missiles and space rockets. The rockets they developed grew ever larger and more powerful. Each of the five Rocketdyne F-1 engines, the most powerful engines ever produced, and which propelled the first stage of the Saturn moon rocket, had a 55,000-horsepower turbopump. It drove 15,000 gallons (57,000 litres) of RP-1 fuel and 25,000 gallons (95,000 litres) of oxygen per minute into the combustion chamber, enabling each engine to produce 1.5 million pounds of thrust. Jeff Bezos and Elon Musk are using similar technology to pump rocket fuels to the combustion chambers of their own privately funded rockets.

Machinery using rapidly rotating wheels, fans, impellers, cogs, and rotors have therefore helped us transform our world in the last two hundred years. It has produced our electricity, driven our pumps, powered our motors, and propelled our vehicles. Its success might readily give us a Panglossian sense that it represents the best possible of all technologies. Indeed we tend to assume that it is the wheel that has been the key to all of our technological progress and could never be bettered. Yet as we shall see in the next part of this book, much of the technology we actually use in our everyday lives employs not wheels but arrangements of levers and joints. This technology is very different, as the elements can rotate about not one but several axes. Consequently, though it looks simple and primitive, it can actually prove more subtle and versatile in operation than our conventional machinery. And we all use it without realising it, as it is the technology that we use to power and control the movements of our own bodies.

PART III

SPIN AND THE HUMAN BODY

CHAPTER 14

Standing and Starting

As a boy I was never a fan of Action Man, or G.I. Joe, as he is known in the United States. It wasn't because I thought dolls were just for girls, because I had a massive collection of furry toy animals. Nor was it due to a distaste for the military; I spent huge amounts of time making model fighter planes, bombers, and tanks, and I had several platoons of plastic infantrymen to act out battles. The problem was that, unlike toy hedgehogs, plastic farm animals, or even the infantrymen, which came on large moulded bases, Action Man was barely able to stand up. He might have had gripping hands and formidable weaponry, but how could a boy respect a doll that had to be propped up against the wall? Later in life I noticed that my partner's daughters did not seem to put so much store on stability. They played happily with their top-heavy Barbie dolls for hours. But perhaps this was because Barbie spent so much time sitting on a sofa, chatting with her friends, or lying in bed with Ken.

The instability of dolls, though very much a first-world problem, should give a salutary lesson to students of anthropology; it shows what a great achievement it is for tall, slender mammals such as humans to be able to balance on our small feet. Indeed, the ability to stand upright has long been seen as a key step in the evolution of humanity. It used to be assumed that early species of hominins acquired the ability when they were living on the savanna floor. The famous picture shows an evolutionary line, with knuckle-walking apes gradually transforming into stooped, hairy ape-men and finally into upright-striding humans. It is now becoming clear, however, that our ancestors may well have acquired the ability to walk bipedally when they were still living in trees. The most bipedal of our living relatives is the most arboreal one, the orangutan, and scientists have recently discovered a twelve-million-year-old fossil ape, *Danuvius guggenmosi*, whose femur bent sharply inward at the top; this would have allowed it to walk upright with the legs under the body. Later fossil apes that were clearly along the line that led to humanity, *Orrorin tugenensis* and *Ardipithecus ramidus*, also had legs that would have enabled them to walk upright, even though they also had powerful arms and shoulders and curved fingers and toes that would have helped them move about in the forest canopy. The idea that we learned to walk in the forest canopy is also a sensible one mechanically. An ape could balance itself far more easily as it was learning to walk if it had branches above its head to hang on to with its hands, and if it could stabilise itself by clinging on to the lower branches with its prehensile feet.

By 3.5 million years ago, the fossils tell us that our hominin ancestors had become at least semiterrestrial. Though Australopiths had powerful upper bodies, their lower bodies were strikingly similar to our own; they had arched feet and straight legs. They must have been able to balance and walk around bipedally on open ground, just like us. This would have enabled them to raise their heads above the long grasses of the savannas, improving their ability to spot both prey and predators, and exposing their bodies to the cooling breeze. It enabled them to walk and run in new ways that were more energy-efficient and, as we shall later see, it would also have freed their hands for other purposes: to carry, wield, and make tools; brandish and throw weapons. Once they could walk upright, the way was open for this insignificant ape to spread and become a top predator wherever it went. So any story about human evolution should be able to describe how, unlike Action Man, we manage to stand upright on two small feet.

Textbooks of biomechanics certainly give a superficially plausible account of how we balance. They model us as rigid oversize dolls who are constantly monitoring the position of our centre of mass. We balance by holding our body upright so that our centre of mass is always directly over our feet. As long as we keep it within the plane of support behind our toes and in front of our heels, we are stable, and we can make sure that this is always the case using our ankle muscles. If our centre of mass moves forward and we are in danger of falling on our nose, we simply contract our gastrocnemius muscles, pulling our Achilles tendons upward and extending our ankles so we press down more with our toes. This moves the centre of pressure of our feet farther forward and produces a moment that rotates our bodies backwards again. In contrast, if we are in danger of falling backwards, we relax our gastrocnemius muscles and contract our tibialis anterior muscles, raising the toes and bringing the centre of pressure back to our heels. Balancing left and right is easier because our feet tend to be relatively far apart, but we move the centre of pressure between our left and right legs using the muscles of our hips to tilt our pelvis to apply more or less force with each foot.

This explanation works well for most people most of the time, but it doesn't explain how we regain balance if we cannot use our ankle muscles for some reason. For example, it doesn't explain how ballerinas can stand for several minutes at a time en pointe, how stilt walkers, tightrope walkers, or unicyclists balance, or how people with foot prostheses stand up. It does not even explain how you or I can balance on tiptoe, or even on the toes of one foot. If you have a vanishingly small area of support, and no way of applying a turning force on the ground, you would topple over immediately, like Action Man. You might think you could restore your balance by moving your arms away from your fall to move the centre of mass back above the area of support, but this would not work. Without applying a lateral force to the ground you would not be able to move your centre of mass. Moving your arms in one direction would simply cause your body to move farther in the other direction and you would fall over just the same. You might also think that you could lean your upper body away from the fall, but this would not work, either. If you try standing on your toes or on the toes of a single foot, you will find that you actually balance by swaying at the waist.

In fact, counterintuitively, the best way to avoid falling over is to bend your body at the hips *towards* the direction of the fall. If you are at the edge of a cliff and want to avoid falling forward, you bend forward at the hip; if you want to avoid falling back into a pond, you bend back at the hip. Biomechanics call this the "hip strategy," but strikingly fail to explain why it works. In fact, there are two ways of looking at the problem, the first being to consider angular momentum. Since there is no way of producing a torque around your feet, the angular momentum of your body will be constant. So if you bend forward at the waist, your upper body will tend to have a nose-down rotation, and so overall the centre of mass of your body will move backwards to compensate. The other, clearer, explanation is to look at the forces involved. Bending forward at the waist will tend to pull your feet forward, but since they are planted firmly on the ground this will set up a backwards force that will push your centre of mass backwards. Of course, this force is only temporary. If you stop bending forward, the force will disappear, and if you straighten back up, this will set up a forward force in the opposite direction. The reason why the movement helps you balance is that bending forward moves your centre of mass backwards within your area of support for a long enough time for gravity to pull your body back. Once you are safely balanced, you can slowly straighten your body.

Ballerinas owe their exquisite balance to their flexible hips, combined with core body strength and lightning-fast reflexes. But not all of us are so graceful, especially when we undertake tricky sports such as ice-skating. If we feel we are falling over, many of us resort to swinging our arms about, Keystone Kop fashion. If we feel we are falling backwards, we swing our arms backwards, and swing them forward if we fear we are falling forward. This has the same effect as the ballerina's graceful swaying of the hips, generating a force on our feet that tends to push our bodies back over our area of support. But this mechanism is not so effective as bending our bodies, which

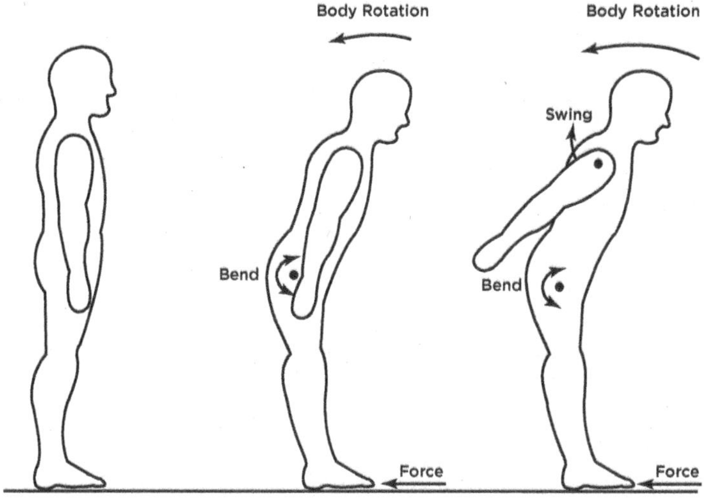

Ways of preventing yourself from falling forward. Counterintuitively, one way is to bend forward at the hips (*centre*), which creates a temporary backward force on your feet, so the centre of mass of the whole body rotates backward. In extreme situations the force can be increased by rotating your arms, like in the butterfly swimming stroke (*right*), and as seen in silent films.

is why in the silent films, the Kops so frequently totter and fall over. Gymnasts who are in real danger of falling from the balance beam do something similar; they raise one leg in the direction of the lean in a desperate attempt to right their balance.

Other animals use similar techniques to balance when they move around in trees or on the ground. Those long-armed primates, the gibbons, walk along branches with their arms held aloft, swinging them back and forth, rather than bending their hips, to balance, a technique that human tight-rope walkers also employ. Quadrupedal mammals such as monkeys and cats have particular problems when they walk along narrow branches, as they have to balance their bodies side to side, not only above their front legs but also above their back legs. To do this, they enlist the help of their tails. As cats walk along a fence, for instance, they hold their tails upright and sway them to the right or left as necessary to keep their hindquarters balanced. The importance of the tail as a balancing organ is demonstrated by the relationship between tail size and habitat preference in mammals; arboreal cats such as leopards have much longer tails than terrestrial cats such as lions; and arboreal monkeys such as langurs have much longer tails than terrestrial monkeys such as baboons. Large bipedal apes like us, meanwhile, can use their hips and arms to balance themselves, which is no doubt why we no longer have any need of a tail.

All of these techniques that involve flexing vertical parts of our bodies from side to side are fairly inefficient because they involve quite large movements of the body *towards* the direction of the fall, which generate quite small forces, and movements of the centre of mass, *away* from it. A much better way of balancing is to use horizontal balancing organs and to move them asymmetrically up and down. A gymnast walking along a balance beam improves her stability by holding her arms straight out to either side. Raising and lowering her arms asymmetrically produces a balancing force much more easily than swaying the hips, and has the additional advantage that the movements can be disguised as graceful gestures. The longer and heavier the arms, the better, which is why tightrope walkers employ long balance poles that they can tilt up and down so they are in little danger of falling. And many bipedal animals have used and continue to use this strategy. Bipedal dinosaurs such as the giant Cretaceous carnivore *Tyrannosaurus rex* and its hapless prey, the duck-billed herbivore *Trachodon*, both held their bodies horizontal, with their long tails held rigidly behind them. They would have been able to balance simply by tilting their bodies slightly, much as a kangaroo or a pheasant does nowadays.

Just as our jointed bodies help us balance, returning our centre of mass back above our feet, so they can also help us to start to move. But setting off into a walk is a surprisingly difficult thing to do. Babies learn how to stand up at the age of around ten months, but they take a couple more months before they start to walk, and often seem to be

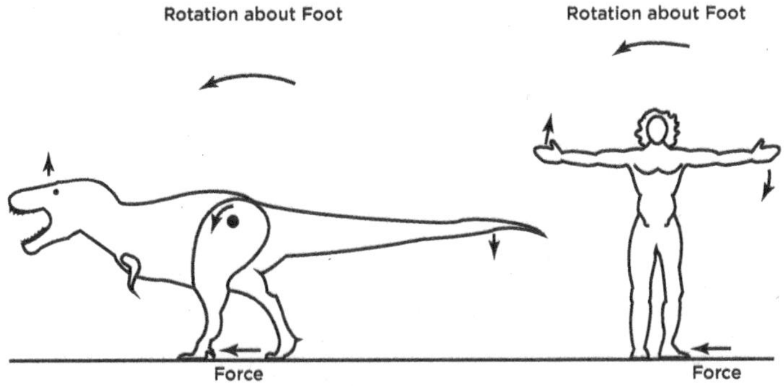

Ways of balancing and accelerating using horizontal-balance organs. T. rex could balance fore and aft and accelerate by swinging its head and tail up and down (*left*), while Leonardo's Vitruvian Man (*right*) would have been able to balance by asymmetrically raising and lowering his arms.

rooted to the spot. The same is true of another group. Elderly people who have suffered from a series of minor strokes, and are in what is known to doctors as a multi-infarct state, can develop problems with gait initiation. They can walk perfectly well, but have great difficulty setting off. The strokes have damaged the pattern generator that they developed to initiate locomotion decades before when they were toddlers. Twenty years ago, as the only biomechanics expert in our university, I was fortunate to be called on to help supervise a study that sought to investigate how these people could be helped to overcome their difficulties. The study was devised and carried out by a talented physiotherapist, Jane Mickelborough, who sensibly decided that the best way to begin was by examining how healthy elderly people start walking, and then to investigate how her stroke patients were different. It was being involved in this study that first got me thinking seriously about how people move.

You might think it would be a simple matter to start walking. All you would need to do would be to raise one leg, move it forward, and plant it back on the ground in front of you. However, if you tried this you would not move forward but sideways and topple over in the direction of your raised leg. Instead, you need to do two things. First, you need to move the centre of pressure backwards to your heels so that you start to fall forward; and you also need to transfer your centre of mass sideways towards your stance foot so you don't fall over when you raise the other foot. We do all of this without thinking when we start walking, which is fortunate because if you do think about it, it becomes confusing! The first stage is easily achieved; all you need to do is flex your ankles to raise your toes so your weight is shifted to your heels. This is achieved by exactly the same mechanism that we use to prevent ourselves falling backwards, by inactivating our gastrocnemius muscles, and activating the tibialis anterior muscles. The other aspect is a bit trickier. To shift the weight to the stance foot, you first have to apply more pressure on the *swing* foot, the one you are about to lift, and you can do this by tilting your hips so that the stance foot is slightly raised. The extra force on the swing foot rotates the body to the other side until the centre of mass lies above the stance foot. At this point you tilt your hips in the opposite direction. The stance foot takes over weight support, allowing you to raise your swing foot and move it forward on its first step. This process is so confusing and counterintuitive that it is not surprising children take so long to learn it! And having lost control of the pattern generator due to the brain damage their strokes had caused, it was not surprising that multi-infarct patients found it so difficult to learn how to start walking again. Many of them developed a technique that involved standing with their legs apart, swaying from side to side, until they could transfer their weight to one foot and start off, taking short steps at first and gradually speeding up and bringing their legs inward. Jane found that giving patients a stick or a bar to hold on to could help them speed up the starting process.

But there is another way that people can use to start to move, even ballerinas who are balancing on their toes, amputees who are standing on their prostheses, or stilt walkers who are standing on their stilts. Just as we can flex our bodies to balance, so can we use

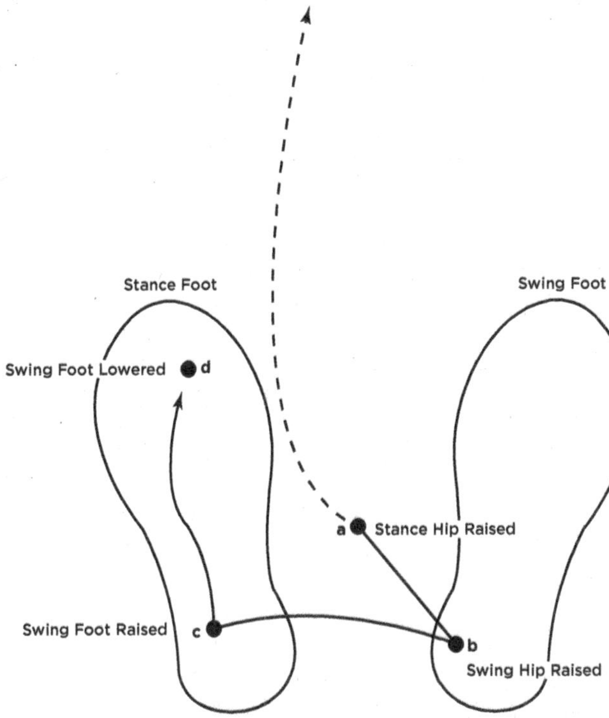

Movements of the centre of pressure (solid line) *and centre of mass* (dashed line) *during gait initiation.* From initial balance (*a*) the toes are raised and the hip tilted to move the centre of pressure to the heel of the swing foot (*b*). This moves the centre of mass forward and toward the stance leg. As the body moves forward, the hip is tilted in the opposite direction to set the centre of pressure to the heel of the stance foot (*c*). At this point the swing foot is raised and moves forward as the body leans forward and the centre of pressure moves forward to the toe of the stance foot (*d*).

the same technique to disrupt our balance and start moving. All we need to do to start moving forward is to bend backwards at the hip and sideways away from our swing leg. We will automatically start on our first stride forward. In the same way bending forward at the hip will set up a force that starts us walking backwards! You can practise this at home, but you can also see it in operation on your televisions. If you watch the ballroom scenes in Jane Austen films (they all have them), gentlemen such as Mr. Darcy and Mr. Knightley extend their backs as they start moving forward at the beginning of a dance, which has the added advantage of showing off their pride and dignity. In contrast, courtiers in historical films walk backwards almost automatically when they bow and scrape before Tudor kings and queens. The comic aspect of Monsieur Hulot's walk in the films of Jacques Tati is that he leans forward as he sets off walking, as if desperate to get going. In fact, this movement moves his centre of mass backwards, putting his weight on his rear foot, so it actually delays his departure—he only starts moving forward when he stands upright again. And Monsieur Hulot was not alone in being slow to get going. Users of the ill-fated Segway often found them hard to stay on as they started moving. As the wheels pushed them forward, the body of the device tended to rotate backwards, depositing the user back onto the floor. And just as our vertical bodies are relatively inefficient at helping us balance, so they are also inefficient at accelerating us. This is why sprinters have long since abandoned beginning races from a standing start, and instead crouch down and use starting blocks so that they can push backwards by extending their legs. The dinosaurs, with their horizontal bodies and tails, would have been able to develop starting forces far more efficiently. They would have been able to accelerate rapidly even without starting blocks simply by tilting their bodies backwards!

But once we have started moving, we have to develop ways of continuing to progress while we plant each foot in turn in front of the other. As we shall see in the next chapter we have learned to do this automatically by exploiting hitherto unexpected properties of our multi-jointed bodies.

CHAPTER 15

Walking and Running

Ever since 1920, when the Czech writer Karel Čapek introduced robots to the world in his play *R.U.R.*, the idea of mechanical humanoid slaves has loomed large in our cultural imagination. However, despite this, engineers and computer scientists have been conspicuously unsuccessful in transforming fiction into reality. The toy robots of my childhood, for instance, were slow clumsy machines. They walked painfully slowly with extremely short steps and they had huge overlapping feet that they needed in order to stop themselves from toppling over. The complexities of the multi-jointed human body have meant that until recently technology has failed to make a robot that could stand up and move around with the fluidity of a real human being. So in Hollywood films like *Star Wars*, the robots were invariably played by actors in robot suits. In that film, the more humanlike robot, C-3PO, like so many robots before and since, walked about in a stereotypical "robotic" fashion, with short, jerky motions. There was never any indication that he would be able to run.

In more recent years, improvements in computation have enabled scientists to build robots that are capable of more lifelike movements, the best-known example being Honda's Asimo. This half-scale robot was programmed to precisely mimic human walking movements. This certainly enabled him to walk around like a person, albeit in a rather furtive way, but he used over twenty times more energy to move a set distance than a real person, and he had none of an athlete's loose-limbed grace. One of the problems of robots such as Asimo is that his engineers treated the body as a series of separate-jointed limbs, each of which has to be powered and controlled individually; they did not appreciate that the movements of one joint in our bodies can actively power and control distant limb segments. More recently, a small group of robotics experts have started to show that walking and running can be powered by very few muscles and are in fact relatively straightforward processes that demand very little in the way of active control. Moreover, they can rival and surpass wheeled transport in terms of economy. After all, as Wordsworth demonstrated, we can think poetic thoughts as we wander lonely as a cloud among the Lakeland fells, or we can make groundbreaking scientific discoveries like Darwin as he walked along his "thinking path." We can do this because the process of walking is largely automatic.

For the purposes of explaining the mechanics of human walking, the late great expert on animal locomotion, Robert McNeill Alexander, showed that we can model our legs as two rigid inverted pendulums. When we walk, we pole-vault over each leg in turn, keeping our stance leg straight, with our heel firmly planted on the ground so that our hips move forward in a series of shallow circular arcs. Since our bodies carry on moving forward, we therefore save most of the kinetic energy in our bodies from one step to the next. However, as we plant each foot on the floor, it does slow our body down a little and diverts its motion upward, so we need to input a small amount of energy at each step to keep moving forward; we push down on the ball of our foot just before we plant the other foot down, which raises our body and rotates it forward over the other leg. To do this, we only need to activate one muscle, our gastrocnemius; it pulls upward on our Achilles tendon and lowers our foot.

But, of course, that is not the only process that happens in walking. We also lift and bend the swing leg, rotating it forward and then straightening it as the stance leg rotates back, and we also swing both of our arms back and forth in time with the movements of the leg on the opposite side of our bodies. It's tempting to think that these processes also need to be powered and actively controlled. However, the design of robots that are based on some of the simplest of children's toys suggests otherwise.

I have fond memories as a child of a walking toy, a plastic representation of Fred Flintstone and Barney Rubble, walking one behind the other. The model had two pairs of short, freely rotating legs that ended in large feet, so that if you set the toy at the top of a ramp it would walk downhill, swaying automatically from side to side as it did so, which raised the swing feet and let them move forward under the influence of gravity. Apart from the lack of knees, the motion was surprisingly lifelike. The success of toys such as these persuaded the American engineer Tad McGeer of Simon Fraser University to experiment with simple "passive dynamic" robots with jointed legs. Working in the late 1980s, McGeer made a simple 2D walker that had four legs, a pair on each side, all rotating about a common hip joint and joined together at the feet; one pair swung inside the other so that the model was laterally stable and could not fall over sideways. Each leg had a knee halfway down it that, like our own, could bend backwards, but that locked straight when the lower leg was bent forward. When McGeer set his model on a slope, it walked down it with leg motions strikingly similar to our own, bending and then straightening the swing legs as they moved forward. And this happened even though there were no muscles to power the movements of the hips or knees. In the early years of this century, Steven Collins of Cornell University combined aspects of the two models to make a 3D robot with jointed legs. This successfully walked down ramps as shallow as 3 degrees, swaying from side to side slightly, like my Fred Flintstone toy, and bending and straightening its swing leg like McGeer's robot in just the way that humans do when walking. He was even able to show that this ability was not just confined to robots that were walking passively down slopes. He added a simple power control unit to his robots that sensed when one of its feet touched the ground, and rotated its other foot downward, just as the gastrocnemius muscle operates in a real person. These robots could actually walk on level ground, and Collins produced some fascinating videos of his robots strolling along a corridor in Cornell. His powered robots were not only able to walk just like us, but they also used a very similar amount of energy; they were twenty times as efficient as the complex computer-controlled robot Asimo.

But how could a robot bend its legs and swing them forward without any muscles? The answer is that the power comes from the other leg, the stance, as it takes the weight of the robot and accelerates its body upward. The obvious parallel to this action is that of a pendulum as shown in the diagram below. In a simple pendulum the weight of the bob accelerates it downward towards the bottom of the swing. But there is another way of speeding up the pendulum or even taking over the accelerating force from the weight of the bob. Its base can be accelerated upward. In the walking robots, the peak force on the stance leg is greater than the weight of the body, so the robot's hips are accelerated upward at the start of the step. This accelerates the swing leg faster than if it were simply raised and released when the robot was stationary. I call this "enhanced pendulum action." And there is another way that a distal leg segment can be accelerated. If the inner segment is rotating, its end will be accelerating inward, so this will accelerate the distal leg segment forward, a process that I call "sling action."

The action of a leg that can bend as it swings forward is slightly more complex than one made of a single rigid segment. It acts as what is called a compound pendulum, which involves both enhanced pendulum and sling action. At the start of a step when the leg is hanging and bent backwards, enhanced pendulum action will accelerate both parts of the leg forward, especially the thigh, so the knee initially bends a bit more. As the thigh rotates, however, the knee accelerates upward and so provides a sling action that rapidly accelerates the lower leg. Halfway through the stance phase, the thigh passes by the vertical and starts to decelerate, like a pendulum coming up the far side of its swing. By the end of the stance phase, the swing leg has automatically straightened and is just starting to swing backwards, so it lands squarely on the ground. Our leg swing therefore happens automatically, and it even adjusts itself to the cadence of our stride. If

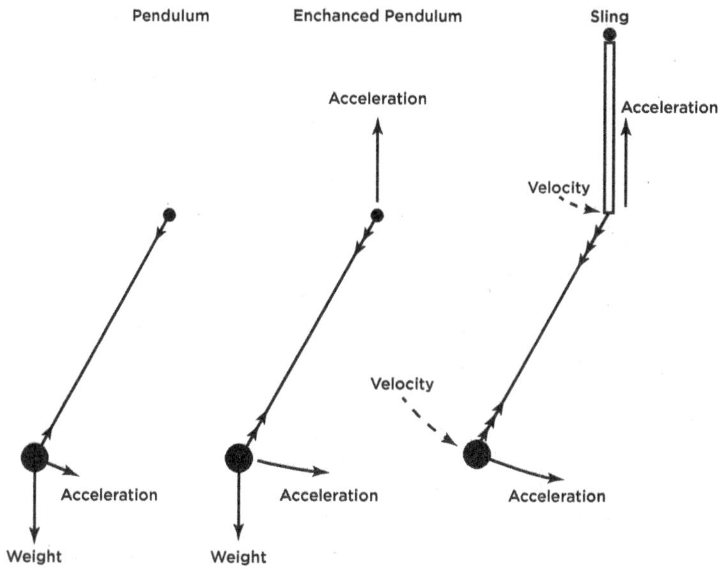

The ways of accelerating a weight without using muscles at the base. The easiest way is to use gravity, allowing it to swing down like a pendulum (*left*). Pendulum action can also be enhanced by accelerating the base of the joint upward (*centre*), as happens at the beginning of the step in walking. A final mechanism is to rotate the basal segment so it accelerates upward and imparts a sling action to the weight (*right*), as when we straighten our knees during the middle of a walking stride.

we walk faster, the stance leg hits the ground harder and the ground reaction force rises to a higher peak, which automatically propels the swing leg forward more quickly.

But our legs are not the only parts of our bodies that move when we walk along; our arms do as well, swinging 180 degrees out of phase with each other and with the legs on the same side. Steven Collins was also able to reproduce this motion in his robots. He simply hung metal wires from the edges of the hip region of his robots (they had no bodies, which simplified balance), and the arms automatically swung back and forth just like our own arms. The source of our arm swing is not hard to fathom. Our shoulders are accelerated up and down as

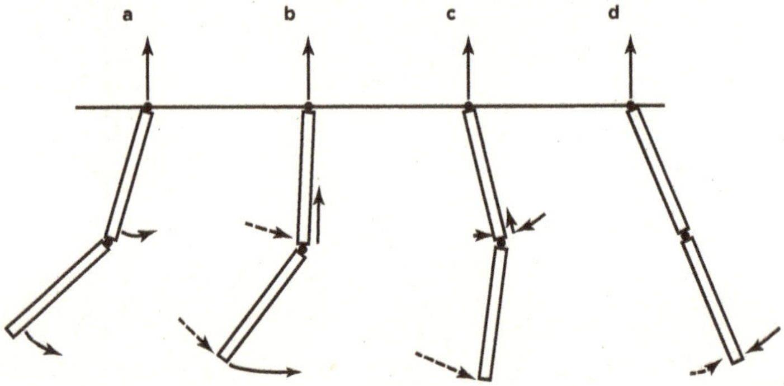

Movements of a compound pendulum such as a limb under the influence of enhanced gravity. The two segments accelerate (*solid lines*) (a) to similar velocities (*dashed lines*) (b). The inward acceleration of the upper segment then accelerates the lower segment faster by sling action. By (c) the upper segment is decelerating, and by (d) it is stationary as the lower segment also reaches the top of its through swing.

we walk along, just like our hips, so our arms also act like compound pendulums, swinging under enhanced gravity, just like our legs. However, it is slightly more difficult to understand why we swing our arms out of phase with the legs, rather than swinging them in phase with the legs or swinging them in phase with each other.

It's certainly possible to walk along with our arms hanging limply by our sides, or to swing them in time with each other, or in phase with the leg on the same side. Back in the 1980s, when I was a chorus member in a production of Gilbert and Sullivan's comic opera *Ruddigore*, we had to march onto the stage, and we were introduced by two members of the cast, who were army cadets and thus experts on marching, to the dangers of ticktocking. This was long before the days of social media, of course, so they were referring not to the production of amateur videos, but to the perils of swinging the arms in the wrong way. When you set off marching, it's all too easy to swing your arms in the same direction as the leg on the same side, and once you've started to march like this it is incredibly hard to get back to normal. The combined movements of the arms and legs unbalance the body and you start to lurch from side to side and can even fall over. And even if you manage to keep upright, you have to exert large torque forces on the ground to rotate your limbs back and forth, which increases the energy you need to walk.

In normal walking, though, the movements of our arms are synchronised automatically by a sling action. As we swing our legs back and forth, the reaction moment swings our upper body in the opposite direction, twisting our back and rotating our shoulders. It is this movement that generates a gentle sling action that sets our arms swinging with the shoulders, and that, together with the vertical acceleration of the shoulders, powers the action. The normal arm swing also has benefits, as Collins and his group showed. They carried out a series of experiments in which people were set walking either with a normal arm swing, with their arms ticktocking, or with their arms held still or bound to their bodies. The experimenters simultaneously measured their oxygen consumption to measure the metabolic cost of locomotion. They found that swinging the arms normally reduced the cost of walking by around 20 percent, compared to holding them still, whereas ticktocking increased the cost by 60 percent. The reason was that the normal arm swing helps balance the body, halving the overall changes in angular momentum of the body about the vertical axis at every stride. It therefore reduces the energy the leg muscles have to exert to press against the ground to produce the required changes in angular momentum.

The characteristic human walk with a stiff, straight stance leg and a freely swinging and bending swing leg is a highly efficient form of locomotion and appeared early in human evolution. We know that our Australopith ancestors walked in the same way as we do more than three million years ago, because their fossil footprints show that they landed on their heel and pushed off using the ball of the foot just like us. It must have given these early humans an advantage when it came to moving effortlessly though the savanna and following herds of antelope. But it is also possible to walk without landing on our heels and pushing off with the front of our feet to power the new stride. Ballerinas, stilt walkers, and amputees can all still walk, even though they only touch

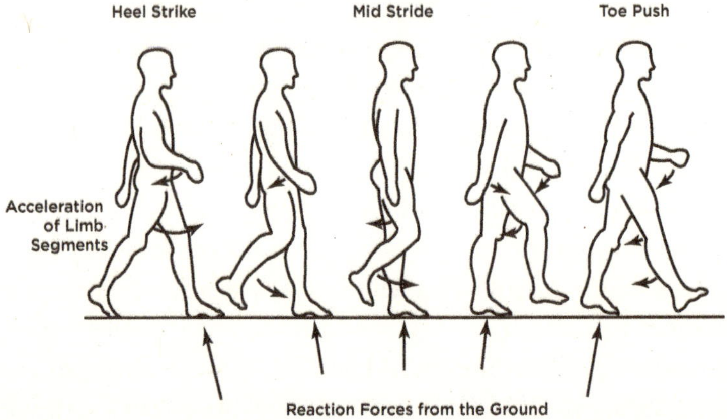

The mechanics of walking. As we pole-vault over our legs, the ground reaction forces are greatest at heel strike (*left*) and on toe push (*right*). In between, the forces are lower. The ground reaction forces accelerate the thigh of the swing leg and upper arm (arrows) back and forth by pendulum action, while the lower leg and forearms are accelerated by a combination of pendulum and sling action. Note that the swing leg is being accelerated backward before it is put down, so it hits the ground vertically. No activation of the leg muscles is needed to move the swing leg.

the ground at a single point. They just have to produce the force that pole-vaults the body over the legs in a different way. Ballerinas walk rather like an ordinary person does when they are tiptoeing along trying to move stealthily. They keep their stance legs bent in the middle of each stride, only straightening them at the end of each step to provide the impulse to move forward into the next. It works, but it wastes a great deal of energy, since the leg muscles have to work hard to keep the bent legs rigid in the middle of the step, which is one reason why ballet dancing is such an exhausting activity. Stilt walkers have an even greater difficulty in moving forward, since their stilts are rigid. To power the start of each stride they have to bend their bodies backwards just like a person trying not to fall over backwards, and then lean forward again later in the stride to reset the body for the next step. It's a strategy the

Dutch robot builder Martijn Wisse of Delft University of Technology used to power his own robot. This arrangement has the disadvantage that because the legs hit the ground harder it is less economical than normal human walking and it also involves large fore and aft swings of the upper body. Birds, which have three main segments to their legs, rather than the two of humans, seem to use a combination of the ballerina and stilt walker technique to power their walk, which is why pigeons nod their heads back and forth at each step.

And there is a limit to how fast even able-bodied people can walk. At speeds over about five miles per hour, gravity is simply not strong enough to keep our feet on the ground as we pole-vault over our legs. If we try to walk faster, we take off and both feet leave the ground, something that is forbidden in the sport of racewalking, which is why racewalkers have to swing their hips in such an exaggerated manner; it keeps their centre of gravity lower in the middle of each stride, flattening the circular path of their hips, and reducing the upward centrifugal force on their bodies. They can therefore move faster before their stance feet lift off the ground. To avoid ridicule, though, the rest of us simply change to a totally different gait at higher speeds: running.

At first glance, running looks very similar to walking. In both modes of locomotion we move along by swinging our legs back and forth and pushing off when our legs are in contact with the ground, and in both we swing our arms out of phase with our legs. But there are also big differences. Unlike walking, running has two distinct phases: the flight phase, when we are sailing through the air, accelerating downward under the influence of gravity, and the stance phase, when one foot is on the ground, at first decelerating the body in both the vertical and horizontal directions, and then accelerating it back upward and forward. The changes in energy that are involved are very different, too. In walking, the kinetic energy of the body is exchanged for gravitational potential energy in each step and back again as the walker vaults up and over the stance leg. In running, both kinetic and gravitational potential energy are lost in the middle of the stance phase and have to be returned. Biomechanics liken running, therefore, not to pole-vaulting, but to the bouncing of a rubber ball or to a person on a pogo stick. To run more efficiently, an animal needs to be able to store the energy that is lost early in the stance phase elastically and release it again. We store this energy in our Achilles tendons and in the ligaments that hold up the arch of our feet; they stretch and shorten again, saving over 50 percent of the energy that we would otherwise need to expend to contract our leg muscles to push off in each step, and ensuring that running is just as efficient as walking.

These differences in the movements of walking and running also result in large quantitative differences in the forces involved. When we are running, our feet are in contact with the ground for much less of the time, usually only between 30 to 40 percent, so the ground reaction forces on our feet are far higher; they can rise to well over three times body weight. These forces consequently produce a far more powerful enhanced pendulum action, which pulls the swing leg more rapidly forward during the stance phase. In contrast, during the flight phase, the body is in free fall, so there will be no pendulum action at all; the legs will continue to swing. For this reason, before we place our swing foot on the ground, we have to actively decelerate it and actively accelerate it backwards again relative to our body; we have to activate the muscles and tendons of our hamstrings and power up the largest muscle in our body, the gluteus maximus. The same is true of the former stance leg; we have to stop it rotating backwards and rotate it forward, using our quadriceps muscles and their tendons. The result is that we have to use our leg muscles far more during running than when walking, and it is during sprints that we can overstretch them, particularly our hamstrings, resulting in damaging muscle pulls.

But the biggest difference between walking and running is in the importance of our arm movements. In walking, our arms move more or less passively, and we can choose to move them in different ways or even hold them still. Because both of our feet are planted flat on the ground for much of each stride, we can easily generate the torque around the vertical axis we need to swing our legs back and forth. In running, only one foot is ever in contact with the ground at any one

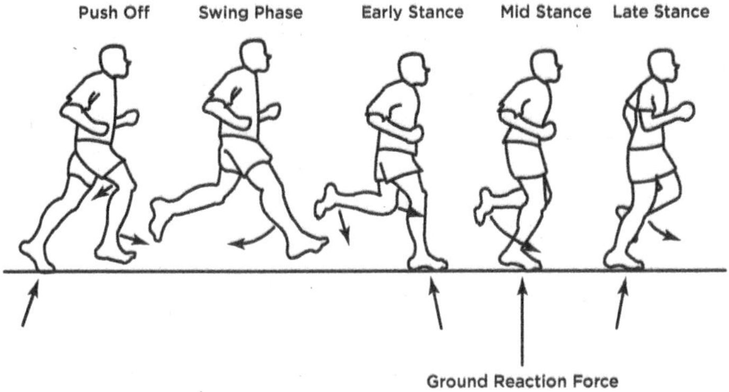

Forces and accelerations of the legs during running. The ground reaction forces power the accelerations of the swing leg (arrows) during the stance phase, with a combination of pendulum and sling action. However, at the end of the swing phase (*second left*) the leg has to be accelerated backward using the gluteal and hamstring muscles.

time, and for most of that time only the ball of the foot touches the ground. There is consequently no way that the foot could alter the total angular momentum of the body about its vertical axis. For this reason, to swing the legs back and forth at each stride we have to swing the rest of our bodies in the opposite direction, and the organs that are best positioned to do this are our arms.

We swing our arms in the opposite direction to the legs, and since they are much lighter we have to move them farther away from the body. It's easy to see just how important arm motion is for running. Try to run holding your arms next to your body and you'll find that it is very difficult; your body twists wildly from side to side and it is very hard to run quickly. Try to run while you ticktock, with the arms moving in the same direction as the legs, and you will find that it is impossible! As we run, our bodies automatically twist in the opposite direction to our leg motion, and this helps amplify the enhanced pendulum action of our arms with a powerful slingshot action. The arm action of runners is best seen in sprinters, and if you watch those overhead shots of Olympic 100-metre races, you will clearly see the opposing motions of the arms and legs. Some of the movement must be powered by the muscles of the torso, which is probably why sprinters spend so much time in the gym building up their upper bodies. However, it is also likely that we store some of the energy needed to swing our arms and legs in elastic structures in the back such as the thoracolumbar fascia and intervertebral discs, though as far as I am aware no one has attempted to measure how much energy is saved in this way.

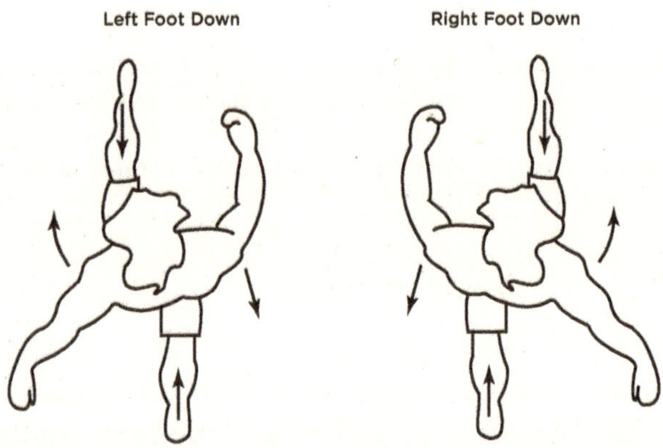

Running from above, showing the opposite motions of the arms and legs that keeps the angular momentum about a vertical axis constant. Note how the shoulders power the arms with a slingshot action.

The final difference between walking and running is in how easy they are for us to control. Because of the short time our legs are in contact with the ground and the small area of contact, it is hard to change direction or adjust our balance when running. We have to be very careful where we put our feet and we are far more likely to trip over or turn our ankles than when walking. It is why so many of us passed our childhoods with perpetually grazed and scabby knees. It's why so few poets produce memorable verse or scientists devise groundbreaking theories while they are out on their morning run. And it's why building running robots is a far more difficult proposition than ones that can only walk. It is only recently that Boston Dynamics has produced robots with a sophisticated-enough computer control system so that they can not only walk but also run across rough ground. But this demands huge computing power, so Hollywood continues to use CG-enhanced human actors to play macho male robots such as *Blade Runner 2049*'s cyborgs, and those high-concept female robots with whom male leads are all too prone to fall in love.

Despite the added complications of running, it is plain that moving about using our complex jointed bodies is a far more straightforward and automatic process than scientists have generally supposed. Together, the enhanced pendulum effect and slingshot effect power and coordinate the swing of our legs and arms, allowing us to make forward progress while balancing our top-heavy bodies. They also help us move at a surprisingly low metabolic and computational cost, which many anthropologists believe could have improved the endurance of early human hunters and helped them run down their prey. And our jointed bodies also allow us to move fairly easily over rough or sloping ground. We just need to make the small adjustments to move our legs slightly differently, which has the only disadvantage that we have to make far more use of our leg muscles. It's why walking over rough terrain can use up twice the energy of walking on flat, level ground, and leaves us with such tired and leaden legs. Nevertheless, our jointed bodies equip us superbly to walk even over terrain that would be impassable to a wheeled vehicle, hence the continued popularity of hill walking and running; we can walk just about anywhere and there is no need for the expensive trails that have to be constructed and maintained for mountain biking. Meanwhile, as we shall see in the next two chapters, we have learned to use our arms, freed up by our bipedal gait, for far more than balancing our bodies when we walk and run. We can also swing them about, using techniques similar to those we use in walking and running, to transfer power from our bodies to our hands, to wield tools and throw weapons.

CHAPTER 16

Hitting

Some of the most disturbing sequences in Stanley Kubrick's 1968 masterpiece, *2001: A Space Odyssey*, are the opening scenes, in which a band of hominins on the African plains learn how to use a tool—a bone club. Given our society's obsession with violence, it is perhaps inevitable that this tool was quickly used for warfare: to kill and maim members of a rival band of hominins. Most horrible of all is the final sequence in which an ape-man uses a femur to make a frenzied assault on the skull of an antelope. The most famous shot of all is the final one. The ape-man releases the club and it flies tumbling into the air, upon which it transforms itself into a rotating spacecraft. This first section of the film therefore sets the scene for the entire film, implying that the whole of human history has been driven by our primeval bloodlust.

Kubrick's is a deeply depressing view of humanity, but it is in some ways fairly accurate. For a start, we do wield a huge number of our tools like clubs: hammers, axes, tennis rackets, and the like. It is also likely that clubs were indeed among the first tools our ancestors invented, though they would have been made not of bone but of wood. For the apes from whom we are descended lived not on arid plains, as anthropologists used to think, but in forests and wooded savannas. The great apes all make a host of wooden tools: wooden nests to sleep in; wooden foraging tools such as sticks to extract termites from their mounds and levers to break open seed pods and bees' nests; wooden digging sticks to uproot roots and bulbs; and even wooden spears that they use to probe into hollows of trees and kill bush babies. And our closest relatives, the chimpanzees, also use lumps of both wood and stone as simple clubs to break open nuts. They position the nut into a hollow on the lateral root of a tree and bring a heavy lump of wood or stone down onto it using an overarm action. With their arms raised in front of their bodies, they lower their upper arm, accelerating their elbows downward and inward, and so develop a sling action in their forearms, which accelerates rapidly forward and downward, the motion being helped by the weight of the tool.

Early humans must have done much the same when they started to use wooden and stone tools for hunting and processing food. They could use exactly the same action when using a short spear to stab prey, to pound the earth with a digging stick, or to use heavy rocks to break open the long bones of carcasses and extract the marrow. However, whereas the shoulders of chimps are sloping, the early humans developed broader shoulders. And once our ancestors started to walk and run, they developed the long flexible back that allowed them to swing their upper bodies and arms back and forth, balancing themselves as they strode along, as we saw in the last chapter. This extra flexibility of the torso would have allowed them to add an extra stage to the clubbing process. When we wield a rock, the motion starts not with our shoulder but with our bodies. Starting with our back twisted, and both our shoulders and elbows flexed, we rotate our shoulders using our body muscles. This produces a sling action that rapidly accelerates the upper arm, and the rotation of the upper arm in turn creates a sling action that accelerates the forearm. We therefore use a two-stage sling action, which can be much more powerful than the single-stage one used by chimps. The whole transfer of energy from our body core to the extremities also needs little involvement of the muscles of the arm or wrist. All that we need to do is keep them rigid at the start of the motion and release each joint at the correct moment when sling action is starting to extend them. This is why we can deliver a far more powerful blow than a chimp, even though we have relatively more slender, weaker arms.

Fairly soon, our ancestors were able to increase the power of their blows even further by a simple improvement to their clubbing action. By around 4 million years ago, hominins such as *Australopithecus afarensis* had evolved longer thumbs, which would have enabled them to hold on to objects with a stronger "power grip." Usually this is seen as an adaptation to hold stone tools. However, the hominin thumb lengthened well before the earliest stone tools were made, 3.3 million years ago. So instead, one of the first benefits of the power grip would have been that they could hold on to one end of a stick and use it as an extra limb segment. They could use their shoulder turn to power the swing of their upper arms, which in turn would power the swing of their forearms, which would finally power the swing of the stick. The three-stage sling action would be even more powerful than the two-stage process and would accelerate the tip of the stick to high velocities. Given that *Australopithecus afarensis* also had a very similar upper-body plan and hand morphology to our own, it is extremely likely this hominin had learned how to make these small adjustments. This would have allowed them not only to use stones to break into nuts and to smash open bones, but also to use clubs to hunt and for intergroup warfare, just as Kubrick depicted in his film. Early hominins probably even learned to use bones, which, being made of a denser material than wood, would have made even more effective weapons than simple sticks.

Over time, hominins would have learned to modify and improve their tools and weapons. Wooden clubs are notoriously blunt instruments, and their effectiveness is further reduced because wood is a relatively light material, so over human history people have modified them in many ways to make them inflict more devastating blows. Adding weight to a club, by making it thicker towards the tip, slows down the speed at which we can swing it. But since this gives our muscles a longer period of time to propel them and we use up less

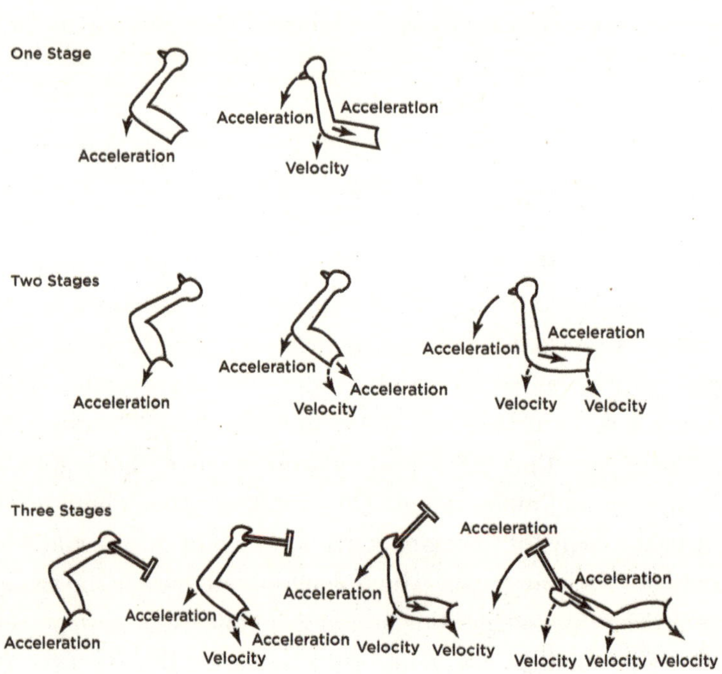

Improvements in hammering power. In one-stage hammering (*top*), moving the forearm down accelerates the elbow inward, accelerating the hand. In two-stage hammering (*middle*), the movement is initiated by accelerating the shoulder, while in three-stage hammering (*bottom*), the rotation of the hand finally accelerates the head of the hammer.

energy accelerating our hands and arms, a heavy club can deliver a more powerful blow. Adding sharp projections to its head makes a club even more effective, because this concentrates the energy of the blow into a smaller area. The club can inflict more damage on the target and might even be able to cut right through it. It is even better to mount a head made of stone onto the end of the club; it is not only heavier but also easier to shape it into a sharp point or blade. The only problem is that it is difficult to attach a stone blade onto a wooden club firmly enough to withstand percussive blows without it falling off or splitting the handle. So it was not until the Mesolithic period, around fifteen thousand years ago, that hafted stone axes became common.

In Mesolithic tranchet axes the flint blade was mounted into a hole that had been drilled through the end of the handle; these axes were powerful enough to cut through the stems of tree saplings, to clear forest glades for hunting, and to allow the hunters to build small lightweight huts. But it was only around eight thousand years ago that the first Neolithic farmers found ways to mount heavy polished stone blades onto wooden handles to make axes and adzes that could cut through a whole tree trunk. This enabled the first farmers in Europe and Asia to clear the land of broad-leaved forests so that they could plant their crops. They went on to use these axes and adzes to build longhouses and roundhouses, and to coppice woodland so that they could use the resprouting poles as firewood or as the raw material to make fences and tool handles. Once the land was cleared, early farmers used mattocks—tools that resemble large-bladed adzes—to break up the soil and enable them to plant rows of seeds. By the end of the Neolithic period, therefore, the vast majority of stone-headed tools were ones that were wielded like clubs and were used for peaceful, productive purposes.

By six thousand years ago, people in Eastern Europe and the Near East had learned how to smelt copper, and shortly afterward, bronze: metals that could be moulded or beaten into sharper blades that for the first time could cut wood across the grain. Bronze was followed two thousand years later by an even harder, tougher metal: iron. Bronze could be cast into razor-sharp axes, adzes, and chisels that carpenters could use to cut the precision joints they needed to construct the first wheels and the first plank ships, while Iron Age smiths learned to use metal hammers to beat hot iron into shape to produce an even wider range of tools. Hand-wielded slingshot tools continued to be dominant in the construction industry for thousands of years, indeed until long after the start of the Industrial Revolution. The great canals, roads, and railways of the eighteenth and nineteenth centuries were built by thousands of "navvies," who broke stones with huge sledgehammers and dug into rock and soil with pickaxes. The balloon-framed homes that housed the Americans who tamed the Midwest were joined together by nails hammered into the laths and boards. Indeed, even today, most craftspeople make extensive use of hammers and mallets.

Not all the innovation went into producing better heads for simple slingshot tools. As we have seen, the more elements there are in a slingshot device, the more effective it is and the faster it can be made to move. To exploit this fact, arable farmers developed a new tool, the flail. This is simply a long-handled stick with a shorter rod tied to its end, which gave a fourth stage to the sling process. Agricultural workers used the flails to thresh grain, to separate the wheat from the chaff. They laid the ears of their harvested cereals in a pile on the floor of an open barn and hit them with the head of the flail, breaking the grains off the stalk and freeing the seed coats, which would blow away in the wind. Threshing was perhaps the most labour-intensive of all agricultural activities; medieval peasants devoted around a quarter of their labour to it, and it occupied most of the winter months. Pastoralists developed a tool that could be moved even faster: the whip. This is a rod with a flexible cord attached at the end. A whipcord can be made to taper, so it consists essentially of an infinite number of increasingly light elements, which can be accelerated in turn by sling action. Indeed there are so many stages to the slingshot action that an experienced user can accelerate the tip of a whip to supersonic speeds, producing the whip crack with which they startle their herds of cattle to control their movements.

Given their ability to transfer energy from the core muscles of a person's body to a fast-moving and destructive point, it is not surprising that slingshot tools were continually being co-opted as weapons to kill other people, just as Kubrick's *2001* depicts. Stone Age societies in Polynesia and New Zealand fought using highly decorated wooden war clubs almost up until living memory, and all around the world people have developed a dizzying array of battle-axes, hammers, and maces. But the most universally feared weapon must have been the sword, a weapon that is basically a flat club with sharp edges. Swords are lighter than most other weapons and since they don't need a heavy head, they are easier to brandish and can be swung about faster to outfight opponents. Swords are also a particularly devastating weapon because a soldier can not only inflict horrendous damage to their adversary using a slashing sling action but also finish them off with a quick thrusting motion. Swords are such effective short-range weapons that they have been taken up and customised by a huge range of "civilizations." In the New World, where people never developed metallurgy, swords were made of wood and were armed along the edges with razor-sharp blades of obsidian. In contrast, in the Old World, swords were forged from steel using the most sophisticated metallurgical techniques known. They ranged from the curved scimitars of Arabia to the heavy two-handed swords of medieval Europe to the exquisite samurai swords of Japan.

But though the lack of a heavy head makes swords easier to swing, it does have one downside. If you slash someone near the tip of your sword, not all of its rotational kinetic energy will be used to harm your opponent. The blade will have an appreciable angular momentum about the point of contact, and will tend to swing forward in your hand, imparting a sharp forward jolt to your wrist. The sword may even be jerked out of your grasp. For this reason, the tip of a sword is known as its foible, as you can impart only a feeble blow to your adversary. If you strike someone near the hilt of your sword, in contrast, the reverse will happen, and you'll receive a sharp jolt pushing your wrist backwards. In this case, though, you will be expecting to have to withstand such a force, so you are ready to resist it. The base of the sword is therefore known as its forte. It is only if you hit someone around two-thirds of the way along the blade, at what is known as the point of percussion, that there will be no reaction force at all on your hand; you will hardly feel the blow, since all the blade's energy will be used to cut efficiently through flesh. It is a nicety that was clearly well known to the English king Henry VIII. When his queen Anne Boleyn failed to bear him a son, he chose the expert French swordsman Jean Rombaud to behead her. It was a final "act of love" to ensure that her death would be swift and painless.

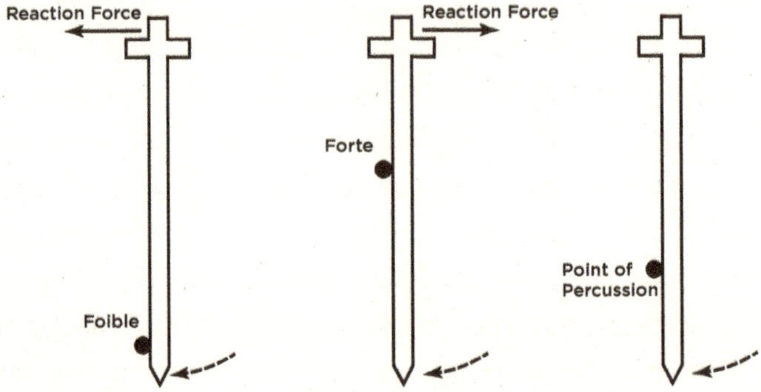

Reaction forces on the hand of a swordsman when he hits an object at different points along his sword. If he connects near the tip (*left*), the reaction force acts forward. If he connects near the base (*centre*), the reaction acts backward. Only if he connects around two-thirds of the way down (*right*) is there no reaction force and the sword sweeps cleanly through.

The key to our success in wielding tools, therefore, is our mastery of rotational motion: we rotate our bodies and shoulders to develop a sling action that accelerates in turn our upper arms and forearms; and we use lengths of rigid materials as extra sling elements to accelerate the heads of our tools to even higher speeds. And as we shall see in the next chapter, we also use this mechanism when we undertake another action that has made us even more feared predators and warriors: throwing.

CHAPTER 17

Throwing

In April 1304, after six years of warfare against the neighbouring Scots, that most warmongering of English kings, Edward I, was on the verge of gaining complete control of Scotland, and laid siege to the last Scottish stronghold, Stirling Castle. But it was a hard fortress to take, since it was built, like Edinburgh Castle, on top of a huge volcanic plug. For four months the Scottish garrison held out against all that Edward could throw against it: lead balls, Greek fire, and even an early form of gunpowder. Infuriated, Edward ordered the construction of a new siege machine; named Warwolf, it was to be the largest trebuchet catapult ever built, taking thirty wagons to transport, but reputedly capable of throwing 300-pound (140-kilogram) rocks hundreds of yards, and capable of destroying any castle at which it was aimed. Petrified, the defenders sued for surrender, but Edward refused to grant it until he had tested his new toy. It was only after four days, in which time Warwolf had destroyed one of the castle's curtain walls, that Edward finally granted the townsmen mercy.

Warwolf was the end result of almost two thousand years of technological development. Trebuchets were first developed in China in the fifth century BC. The early machines had a long wooden beam, hinged a quarter of the way from one end, with a sling mounted at the end of the longer arm, just like later ones, but they were what are known as traction trebuchets: powered by people using their weight to pull downward on the short arm. It was only later that counter-weight trebuchets such as Warwolf were developed, in which people were replaced with a huge weight that was hung from the end of the short arm. These devices were loaded by using a crank or treadwheel to ratchet up the device, giving it more power and greater accuracy. Trebuchets were imported into the Middle East around the fifth century AD and were used as siege weapons by the triumphant Islamic armies, before being adopted in Europe, where they reached their peak of popularity and attained their largest size in the medieval period. They were used not only to attack castles, but also to defend them, being mounted on the tops of the towers. The design and construction of trebuchets were entrusted to skilled craftspeople, and since these huge machines were commonly called engines, their designers were the first people to be known as engineers. They were among the few artisans whose names were recorded and whose fame has passed down through the centuries. Notable among them was Master Bertram of Sault, who first appears in history among the retinue of Henry III of England while on campaign in Gascony. He was later employed by Edward I to design the trebuchets that defended the Tower of London. Warwolf itself was designed by James of St. George, a French architect who had designed and supervised the construction of the many castles Edward I had built in Wales to subdue its inhabitants.

Trebuchets mark the culmination of people's efforts to expand their influence not just by wielding long-handled tools but by actively throwing objects. Hominins had probably been throwing missiles for millions of years: to ward off predators, hunt animals, and kill other people. Yet another of the benefits of standing upright was that it frees our hands and arms, which can be used for this purpose. Other primates, which are mostly quadrupedal, can throw things when they are squatting or standing up, but it is a skill they rarely practise. In contrast, human children spend vast amounts of time learning how to throw and catch balls, honing their hand-eye coordination and skeletomuscular control. The techniques we use to accelerate our arms are broadly the same as those we examined in the last two chapters: those we employ when walking and using hand tools.

The first and most common technique we use to throw objects short distances is the underarm throw. The thrower pulls their arm backwards behind their body, as they would when walking, and then actively swings it forward again before releasing the object. Some of the power is supplied by contracting the muscles in the shoulder and arms, but you can achieve a much more powerful throw if you crouch slightly before the throw and rise again as you start the forward swing, like a Frenchman playing boules. This accelerates the shoulder upward and produces an enhanced pendulum action to accelerate the upper arm forward. The action is demonstrated most clearly by tenpin bowlers, who use it to send bowling balls, which can seem too heavy to even hold, let alone swing using just their arm muscles, at high velocity down the bowling lane. Underarm throwing is easy to learn, and indeed, it is the method used by captive-bred chimps when they throw fruit or stones around their enclosures, but it has its limitations. It allows us to control the direction of the throw fairly accurately, but not the elevation; if we don't time the release accurately the object can scuttle across the ground, or fly vertically into the air. There also seems to be a limit to how far we can throw underarm. To throw more accurately and powerfully, people have to learn an alternative: to throw overarm.

The best way to achieve the most accuracy if you only have to throw an object a short distance is to copy the method that authors in films use to toss the scrunched-up pieces of paper of their aborted novels into their wastepaper baskets. They keep their bodies as still as possible and move just their arms, swinging their forearms in a vertical plane in front of their bodies with the hand starting its motion near their eyes. The method is strikingly similar to the hammering technique employed by chimps that we saw in the last chapter and the overarm technique that some chimps have been taught; it accelerates the forearm by lowering the upper arm and thus powering the forearm with a sling action to straighten the arm. It works reasonably well, the technique being most profitably used by professional darts players, but you cannot throw very fast or very far using this method;

the fastest a powerful chimpanzee can throw overarm is a mere 20 miles per hour (32 kilometres per hour). To put more power into your throw you need to add another stage to the sling action.

The power method that is probably easiest to learn is to throw with your arm out to the side, like when you skim stones. Before you throw, you twist your body and flex your shoulder and elbow so that you can produce a double sling action when you start the throw, rotating your shoulders to power the acceleration of your upper arm, whose rotation in turn powers the acceleration of your forearm. Skilled throwers, however, such as baseball pitchers and fielders in cricket, throw using a proper "overarm" technique with the upper arm slightly above the horizontal and rotating the whole body, starting the action from the legs to power the first stage of the sling process, with the second stage, acceleration of the forearm, following. Throwers can even give a flick of the wrist to add a final, third stage to the throw. Using this technique, baseball pitchers can throw balls at speeds of up to 105 miles per hour (170 kilometres per hour) and cricketers can throw balls distances of up to 300 feet (90 metres).

However, because of the anatomy of our hands and fingers, you can't use the conventional overarm action to throw long slender objects such as spears. Instead, you have to release them with your upper arm almost vertical and with your hand above your head, so you are forced to move your arms in an even more vertical plane. The javelin-throwing technique is hard to master, but a decent practitioner can throw a spear around 100 feet (30 metres), and Olympic athletes are able to throw javelins up to 260 feet (80 metres). The key, once again, is to exploit a double sling action, rotating and lowering the shoulder to accelerate the upper arm, and in turn using its rotation to power a sling action, accelerating the forearm. The spear-throwing action has two further advantages. Moving your arms in a vertical plane makes the direction of your throw more accurate. The accuracy of the technique is also improved by the way you release the spear. As it leaves your hands the handle rolls up your fingers, causing the spear to spin along its length, stabilising it in flight, and allowing it to fly like the fuselage of a plane, straight through the air.

Being able to throw stones and spears tens of yards no doubt gave people a great advantage. It would have helped them fend off predators, hunt game, and fight rival tribes. Since even chimpanzees can throw, hominins must have been able to throw stones from early in their history. And seeing that early hominins had broad shoulders and a flexible waist, just like us, it is possible that they must have been able to throw powerfully, using a double sling action, as early as 3 million years ago. It is also likely that they were able to throw spears, but because wooden implements do not survive in hot tropical regions, we have no evidence to back up this assumption. All we have is evidence, in the form of wear and wood residues on hand axes, that an early member of our own genus, *Homo erectus*, was carving wood 1.5 million years ago. The first undisputable hunting spears were those found in 1994 in Lower Saxony, Germany, which were dated at over 330,000 years old. The Schöningen spears were around 8 feet (2.5 metres) long, beautifully tapered, and thickest towards the front. They closely resemble modern javelins, and experiments have shown that they can be thrown accurately distances of at least 70 feet (20 metres). The spears were found alongside the skeletons of over twenty horses, some of which had wounds showing that they had been killed by a spear. The Neanderthals who had carved the spears were clearly adept throwers.

But good as we are at throwing, people have always been keen to throw even farther, and just as when we wield handheld tools, the best way to improve our throwing performance is to extend our arms, using a further element so that we can produce a triple sling action. The easiest way to do this is to hold a stick not at its centre, like a spear, but at one end, like the handle of a hammer, and to release it at the end of the action, as people do when throwing a stick for their dog. The stick travels faster than the hand at release and tumbles through the air, so it also causes extra damage when it hits a target. People from all around the world have used throwing sticks for hunting, including the Guanches, the original inhabitants of the Canary Islands, and the ancient Egyptians. Egyptian pharaohs such as Tutankhamen were enthusiastic bird hunters, using throwing sticks weighted at the end with a stone tip, as is depicted in many Egyptian tombs. But the best-known throwing sticks must be the boomerangs used by the Aborigines of Australia. Most of these are not curved, but straight lengths of wood, which are carved into a streamlined airfoil cross section to reduce drag and allow them to fly farther. A straight hunting boomerang can be lethal at distances of up to 650 feet (200 metres).

The Aborigines have also learned to exploit the phenomenon of precession we first encountered in chapter 2 by using bent boomerangs that return to their owner. The boomerang is flung with the distal end of the blade pointing forward. As a consequence, the aerodynamic lift that the boomerang produces acts in front and to the outside of its centre of gravity. This generates a turning moment that is converted by the boomerang's rotation into a moment tending to tilt the boomerang upward and inward. It flies in a graceful curve high above its initial flight path and returns falling gently to its owner's hand. It is a technique that is nowadays also exploited in Frisbees, which are just the latest in a long line of throwing disks. These include the chakrams of the Sikhs, which are essentially military equivalents of the plastic throwing rings used by children. They are sharpened bands of steel or brass 6 to 12 inches (15 to 30 centimetres) in diameter. Sikh warriors threw them by twirling them around a finger before sending them spinning at their enemies.

The downside of throwing sticks is that because the hand is travelling more slowly than the tip of the stick, the average speed of the stick at release is relatively low. Together with the large drag on the careering stick, this limits their range. People have consequently developed another series of devices that launch projectiles from their fast-moving tips. Many coarse fishermen use curved "throwing sticks" to propel maggots and other bait into the water around their hook. They place the bait in a cup at the far end of the stick and use an overarm throwing action to accelerate it. Even more people use a similar device to exercise our canine friends: a dog ball thrower. Even poor throwers such as myself can throw tennis balls distances of 100 to 160 feet (30 to 50 metres) with these tools, their effectiveness being boosted by their flexibility, which allows them to store elastic energy in the handle and release it to the ball as it flies out of the cup; there is no need to actively extend the wrist.

The same principle has long been used by hunter-gatherers in the form of "spear-throwers." The atlatls of Central America and the woomeras of Australia are simple rods, some 8 to 12 inches long (20 to 30 centimetres), with a notch or prong on the end, which hooks over the rear of a small spear or large dart. The hunter holds the thrower parallel to the spear shaft, gripping both at the same time, and then goes through a conventional throwing action. The spear-thrower provides the third stage of the sling action, just like a dog ball thrower, and bends the spear at the start of the throw. Energy is therefore stored in the spear, which is released as the spear straightens, increasing the velocity of the spear as it leaves the thrower. The hunter can also control the release and add extra power by actively flexing their wrists, making the device even more deadly. The only downside is that unlike the hands of javelin throwers, spear-throwers do not impart any spin, so Paleolithic hunters had to add flights to the back of their darts, like those on an arrow, to stabilise their flight.

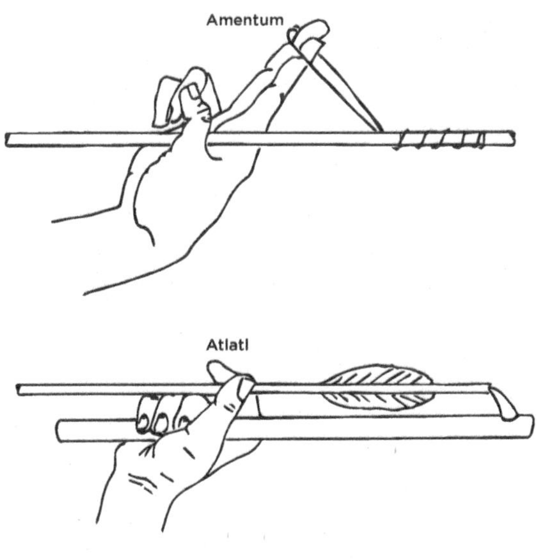

A comparison of the techniques used by ancient Greek javelin throwers (top) *and hunter-gatherers from America and Australia* (bottom). The amentum of the Greeks spins the javelin as well as slings it, but is less accurate, whereas atlatls and woomeras provide more control, but do not spin the dart, which requires a flight.

Since the hunter holds on to both the spear-thrower and the dart at the beginning of the throw, the thrower itself does not even have to be rigid. It can be replaced by a length of cord. In ancient Greece, for example, the lightly armed peltast troops of the fourth century BC threw their javelins using a thong called an amentum, which they looped over two fingers, while they wound the other end around the shaft. They held the middle of the spear with the cord taut, and then threw the spear using a conventional overarm throwing action. By the time the thrower released the spear it would be travelling fast enough for slingshot action to take over. The cord would continue to accelerate the spear before finally unwrapping itself and releasing it. By the time the cord had unwound from the spear it was travelling much faster than the thrower's hand. The process also has a second advantage. As the cord unwraps from the shaft it sets the spear spinning along its length, stabilising its flight and ensuring that the spear hits its target head-on. In more recent times, the miners of South Yorkshire and the peasants of Switzerland used strings looped around a single finger to throw arrows vast distances—up to 650 feet (about 200 metres)—in an unusual competitive sport.

There is even evidence that Upper Paleolithic humans used a similar technique twenty thousand years ago, in the form of a series of Y-shaped rods, carved from antlers or wood, which were unearthed at the turn of the twentieth century, alongside the famous paintings in the caves of Southwest France. Bâtons de commandement, as they are known, are usually 6 to 8 inches (15 to 20 centimetres) long, with a hole drilled into the bifurcation of the main body of the rod into its two short arms. Since they are highly decorated and did not appear to have any obvious use, archaeologists initially assumed that they had a purely ceremonial function like a scepter—hence their name. However, in recent years, devotees of primitive technology, a collection of amateur enthusiasts, have shown that bâtons de commandement were a key component of an apparatus that improved the flight of their spears. The hunter would tie a cord through the hole in the baton and wrap the free end a couple of times around the rear of his spear.

There are good reasons for accepting this interpretation. First of all, one of the batons unearthed from the famous La Madeleine cave in Southwest France has a simple carving showing a man holding a dart just as he would if he was about to throw it with a baton. Some batons also show the sort of wear around the hole that are consistent with this form of use.

And, of course, the proof in the pudding is in the eating; primitive-technology enthusiasts can throw spears up to 200 feet (60 metres) using a baton, twice as far as they could throw it without.

And people have been using the principle of the sling for millennia to throw stones using an even simpler piece of apparatus. The slingshot consists of just a length of cord with a loop at one end and a leather or string pouch halfway along it. A slinger places a stone in the pouch, fits the loop around the forefinger, and holds the other end of the cord between the forefinger and thumb. They then simply twirl the hand in a circular motion, propelling the taut string like a lasso, and release the end of the string when it is at right angles to the movement of the hand so that the stone is released when it is travelling the fastest. I say "simply," but the process is extremely tricky to master. It is hard to accelerate the sling from rest because when the projectile is still it will tend to rotate backwards, rather than forward, if the inner end of the string is accelerated. Slingers have to learn how to get the stone moving while keeping the string taut at all times; unfortunately it's not something I have been able to master when I have tried to use the sling that I bought over the internet!

However, Paleolithic people would have had plenty of time to practise their craft, and they would have passed their knowledge about how to use these powerful weapons on to Neolithic herders, who would have used their slings to protect their animals from wolves and other predators. Indeed, the most famous user of the sling was the shepherd boy David from the Bible, who used his simple weapon to fell the Philistine giant Goliath in just one single blow to the forehead. Slingmen were also commonly seconded to regular military life. They became a feared element of the Roman army, for instance, capable of throwing lethal lead stones over 1,300 feet (400 metres)—farther even than the arrows shot by bowmen. Each torpedo-shaped stone had a bloodthirsty message inscribed on its surface, and some even had

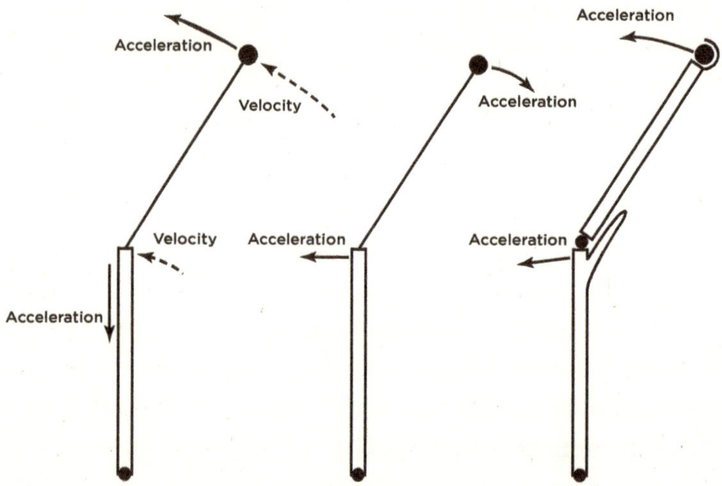

The problem with slingshots. They are efficient in accelerating projectiles when the arm is moving at full speed (*left*). However, at the start, accelerating the arm will cause the projectile to rotate backward (*centre*). This can easily be overcome with a rigid outer arm, which can be held firm at the start of the throw (*right*). Slingers using flexible strings have to use other ways of starting the movement.

small holes drilled through them, which produced an eerie whistling noise as they rained down on the enemy, spreading panic through their ranks. But the slingmen had to be brought up to the task. The crack mercenaries from the Balearic Islands who made up the bulk of the units of slingmen in the Roman army were trained from early childhood in their profession, rather like the English longbowmen of the Middle Ages.

The effectiveness of sling action can easily be demonstrated by the very different performances of athletes in two Olympic throwing events. Hammer throwers, who use a double sling action, rotating several times before giving a final pull on the wire of their hammer, can throw their 16-pound (7.2-kilogram) projectile up to 260 feet (80 metres), imparting around 2,800 joules of energy to it. In contrast, shot-putters, who essentially push their 16-pound (7.2-kilogram) shots, can only throw around 70 feet (21 metres), imparting a mere 720 joules of energy to them.

Having investigated the mechanics of human throwing and of the simple devices people have used over the millennia to improve their performance, we can now better appreciate the design and operation of medieval trebuchets such as Edward I's Warwolf. Modern reconstructions of trebuchets, such as the 60-foot-high (18-metre-high) replica built in 2005 at Warwick Castle in England, show that they would have been truly formidable weapons, and all because they use not one but two of the techniques that humans utilise to throw things. In a trebuchet, the projectile is held not in a cup at the end of the long arm, but in a sling attached to its tip. When it is loaded and wound up, the projectile rests in a groove near to the base of the machine. As the trebuchet arm is released, the short arm with the counterweight falls, accelerating the long arm upward. This accelerates the projectile backwards and outward using the enhanced pendulum action. Once it is free of the ground and the arm is moving at full speed, the projectile is then accelerated upward and forward by the second action: sling action. Finally, at the point when the sling is parallel to the arm of the device, an automatic catch mechanism releases it. Most of the energy from the fall of the counterweight goes into propelling the projectile. If the counterweight is hinged to the end of the beam and free to swing, over 75 percent of its energy is used, and the remainder is gradually lost after the projectile is released as the weight swings more and more slowly backwards and forward. Trebuchets would have been quite safe when they were fired.

The only difficulty with trebuchets is that the behaviour of the whole system is hard to predict, since the forces produced at any moment depend on the current velocities of the arm and the sling, which in turn depend on the situation just previously. These sorts of systems are nonlinear and cannot be analyzed mathematically, which is why physicists have had to resort to carrying out computer simulations of the process. Not having access to computers, the medieval trebuchet engineers relied on years of experience to build up a feel for their devices so that they could construct them and optimise how they were operated to throw stones as far and as accurately as possible.

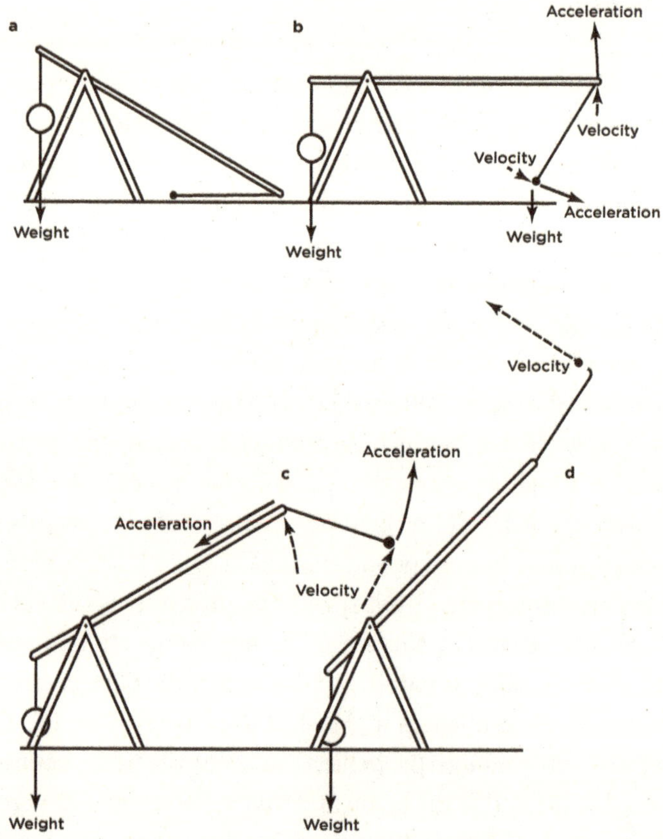

The action of a trebuchet. The projectile is powered by the fall of the counterweight. After release (a), the projectile is first accelerated by an enhanced pendulum effect (b), but as the arm reaches maximum speed (c), the sling effect takes over. The projectile is eventually released when the sling is parallel to the arm (d).

You might think that nowadays we would have little to learn from such a motley collection of slingshot devices. After all, most of them consist merely of a simple stick, a beam, or a piece of string. But these techniques have been exploited again and again throughout history. For instance, bowmen have for ten thousand years used the stabilising effect of spin to improve the flight of their arrows. Fletchers attach the flights at a small angle to the shaft of their arrows, making them rotate as they fly, and enabling them to fly straighter. And since the nineteenth century, the flight of bullets and shells has been stabilised in the same way using rifling. Gunmakers etch a spiral groove into the barrels of guns and cannons, while the makers of bullets and shells design them so that they fit into the grooves and are set spinning as they travel down the barrel. They incorporate a ring of soft metal, the driving band, at the rear of the bullet, which has a slightly larger diameter than the barrel. When the gun is fired, the soft metal of the driving band swages into the grooves to seal the barrel and allow the bullet to move forward and simultaneously spin around ever faster as it accelerates. The introduction of rifling resulted in a dramatic improvement in the accuracy of guns: from the wayward muskets of the seventeenth century, which were just fired in the general direction of the enemy, to the modern high-performance sniper's rifle, capable of firing a bullet with an accuracy of inches over a distance of two miles.

And against all expectations, sling action itself is actually rocket science! The huge rockets that transport astronauts and clusters of satellites into space are stabilised and manoeuvred by small side rockets that expel gases out laterally rather than backwards and that steer them like a rudder. However, it is simply not practicable to use this system to stabilise the huge numbers of small satellites that we put up into orbit. Instead, satellites are stabilised by spinning them along their length like bullets, and at a high rate of rotation—something like fifty revolutions per second. The problem this causes is apparent when the satellites need to get down to their business, however. The spin rate is simply far too fast for them to establish communication with ground control. To overcome these difficulties, in the early 1960s NASA developed a way to slow the spin down. They wrap one or, more usually, two long wires with weights on the end around the body of the satellite and clamp the weights to the side of the craft. When the satellite reaches its final orbit the clamps are opened and the weights are released. The rotation of the satellite gradually swings the weights outward, unwrapping them, and finally flings them forward with a slingshot action and releases them into space. In this process, which NASA dubbed yo-yo de-spin, most of the rotational energy of the craft is transferred to the weights, slowing the rotation of the satellite itself down to a manageable one to two revolutions per second. Since the starting point was known perfectly, NASA mathematicians even managed to analyze this process and publish the (extremely long and complex) equations that describe the motion of the weights and their effectiveness at slowing the spin of the satellite.

It seems that, where throwing is concerned, we still have a lot to learn about ourselves, and from ourselves, and that the ways we throw objects can improve even the most advanced of our technologies. And as we shall see in the next chapter, this is even truer when it comes to the most extreme motions we perform: when we take part in sports.

CHAPTER 18

The Sultans of Spin

Watching elite sports on TV can be a fascinating and addictive pastime—but watching it in the flesh can be even more enjoyable and the exploits of the sportspeople can seem positively magical. As a child I remember being overwhelmed by the balance and athleticism of international cricketers at the Oval, astonished by the ability of the professional tennis players at Wimbledon to hit tennis balls with impossible speed and accuracy, and awed by the power and agility of international rugby players at Twickenham. They all seemed to be capable of superhuman feats. In this chapter we will investigate what it is that makes a great sportsperson, and we will examine the techniques they use to exploit the mechanics of rotational motion to their limits; to move faster, propel projectiles more rapidly and accurately; deceive their opponents; and throw their bodies around in mind-dizzying fashions. As we shall see, it all revolves around their ability to exploit the rotation of the different parts of their bodies.

Opinions often vary about what are the most important attributes sportspeople need to achieve success. Commentators cite many factors: the shapes of their bodies; the power and efficiency of their muscles; the aerobic capacity of their lungs; their dedication to training; and their hand-eye coordination. All must play some part, but a clue to which are most crucial comes from comparing the all-around sporting prowess of stars from different sports. Back in the 1970s, European television ran a series of "Superstars" contests, in which a range of sportsmen (it was all men in those days) were pitted against one another in a series of challenges, from sprinting and cycle races, tennis and Ping-Pong matches, weight lifting and assault courses, to gymnastics. Most of the sportsmen were pretty good at almost all of the events—way above average; the most successful, I recall, was the Austrian ski jumper Karl Schnabl. There was just one exception. Professional cyclists often proved to be hapless at other sports and regularly came last even though they are among the fittest people on the planet.

Why this is the case was demonstrated by a series of experiments carried out by biomechanics and sports scientists. A study by Robbie Wilson of the University of Queensland, Australia, for instance, on a group of semiprofessional football players, investigated which attributes contributed most to their prowess in the sport. To do this, he first measured as many of their physical and sports-specific attributes as he could, from aerobic capacity and muscle power to dribbling skills. He then compared them to their relative overall football-playing ability, as measured by the outcome of one-on-one games between the players. He found that football-playing prowess was more closely correlated to a balance task than to any other metric. Similar results were obtained by Thierry Paillard and his colleagues at the University of Pau, France. They showed that in a range of sports—judo, surfing gymnastics, and football—elite athletes performed better than average ones in balancing on small seesaws; they needed to make much smaller adjustments in the proportion of weight supported by the different feet—less than half as much—during the task, especially when blindfolded.

Thinking about it, these results should not really be surprising. Commentators often mention how well-balanced sportspeople are, and the same is true of great dancers. And as we saw in chapter 15, the techniques we use to balance are also the same as the ones we use to accelerate. Having better balance should enable a person to move more quickly and fluidly and change direction more rapidly; it is an excellent proxy for overall body coordination. Knowing this, therefore, it is not surprising that people who excel in one sport are good at many others. It also explains the relatively poor performance of professional cyclists. As we have seen, bicycles are inherently stable, so cyclists need no special coördination to excel; they just need an exceptional cardiovascular performance and large lungs so that they can apply more power for longer periods to the pedals.

Knowing that most elite sportspeople have better balance and coordination than the rest of us is helpful, but it is only part of the explanation of their sporting prowess. We also need to know more about the mechanics of the techniques that they use so we can appreciate their performance better, make us more informed sports fans, and even help us improve our own sporting performance.

Some of the most popular games worldwide involve kicking a ball—football, rugby, and American and Australian Rules football, for example. Yet kicking is a rather unnatural action given that in nature most objects are so hard that contact with them would damage our naked feet. Sports scientists generally assume that kicks are powered by the muscles in the kicking leg, particularly the quadriceps muscles, which act to extend both the hip and the knee. However, on close examination, it becomes plain that the movements of the kicking leg are very similar to those of the swing leg in walking and especially in running. The mechanics of the kick are consequently also very similar. Anyone aiming to kick something takes a run-up, and there are good reasons why. The obvious one is that the run-up gives an additional velocity to the foot, since the hip is already moving forward. But there is a more important reason; it helps to swing the swing leg forward, just as it does in running itself. As the football player plants their stance foot on the ground, they alter the direction in which the hip is travelling, accelerating it upward. The stance leg therefore powers the swing leg, accelerating the thigh and lower leg with a strong, enhanced pendulum action. As the femur moves forward, it also in turn accelerates the lower leg by a slingshot action. Though the muscles in the swing leg add to the motion, much of the power of the kick therefore comes from the muscles in the stance leg and hips, and the key to kicking a ball rapidly and precisely is not the strength of the leg muscles themselves, but the precise timing of the action. This may be why such slender football players as David Beckham and even short ones such as Lionel Messi can have such powerful shots.

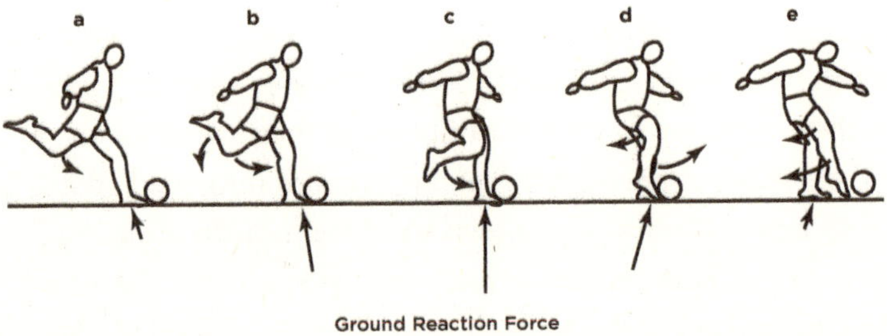

The mechanics of a conventional penalty kick. As the ground reaction force increases forward (a to c) as the stance leg accelerates the hips upward, the femur and then the lower leg are accelerated (*small arrows*) by an enhanced pendulum and sling action. By contact (d) the femur, and after contact (e) both the femur and lower leg, are decelerating.

Apart from the importance of the enhanced pendulum effect in powering the kicking leg, there are two areas in which the mechanics of rotation can further our appreciation of football: taking penalty shots and performing overhead kicks. Penalty-shot taking is the aspect of the game that exposes players to the greatest pressure and that can also reveal flaws in their kicking technique, so it is worthwhile to consider a novel technique that has emerged in the last few years. Most football players run up quite quickly to take a penalty, choosing before they do so which side of the goal to aim at, and kick the ball as hard as possible. The disadvantage of this approach is that if the goalkeeper guesses correctly which direction to dive, they have a reasonable chance of stopping the shot. However, there are two players in the English Premiership who have developed a rather different penalty-taking technique. Both Jorginho of Arsenal and Manchester United's Bruno Fernandes run only slowly up to the penalty spot and then take an almost vertical leap into the air in their final stride before landing and aiming the ball almost unerringly away from the goalkeeper's dive. The hard landing produces a strong enhanced pendulum action that helps power the kick, but it does not make their shots more powerful than conventional penalty takers. The advantage is that with a slower, more controlled approach, these strikers can keep their eyes on the goalkeeper for far longer, and only decide which direction to kick the ball after they have started to move. The technique requires massive skill but can result in success rates of over 90 percent.

If few players can take penalty shots as well as Jorginho and Bruno Fernandes, even fewer are capable of performing overhead bicycle kicks—to shoot at a goal while they are flying upside down through the air. The sheer gymnastic ability to leap in this way is impressive, and the feat is even more impressive when you consider the mechanics of the kicking action. When falling through the air, the football player has no way of powering his swing leg using the enhanced pendulum effect. They have to swing their legs using only their leg muscles, and to stop their bodies rotating in the other direction, they have to swing their other leg backwards and their arms in the opposite directions to the legs. The legs actually cross over in midair, which is why overhead kicks are also known as scissors kicks.

Ball-handling sports such as rugby, American football, and basketball demand another skill set. In these sports, fast-moving players have to be able to avoid being tackled, and they often do this by sidestepping or juking opponents; they appear to be about to pass them on one side before suddenly veering off on the other. Sidestepping involves a modification of the technique that most of us use mainly to accelerate from a standing start, swaying our upper bodies at the hip. In the case of the sidestep, the player leans not forward or backwards, but to one side. This dips one shoulder, so the player seems as if they are starting to move towards the dipped side. However, as we saw in chapter 14, if the feet are on the ground, bending the upper body to one side actually sets up a force that pushes the body in the opposite direction, bamboozling the opponent. Of course, the person who is being sidestepped also knows about the technique of the sidestep, so a skillful sidestepper adds a further level of cunning. As they approach their opponent they sway their upper body rapidly from side to side. The opponent is therefore put in doubt about which direction the player will sidestep, doubt that is made worse because it is so hard to look at both their upper body and their feet at the same time. It is only when their opponent has made a choice about which direction they are likely to turn that the sidestepper puts their foot down on that side and uses their "fast feet" and body sway to accelerate in the opposite direction. Some rugby fly-halves even leap into the air before making their sidestep, just like penalty takers, to maximise the force on their feet and accelerate more rapidly sideways.

Another skill demanded in both rugby and American football is to pass the ellipsoidal ball quickly and accurately to teammates, a skill that is particularly important for scrum-halves and crucial for a quarterback. Both of them use a torpedo pass, spinning the ball around its long axis so that it stays oriented, pointing directly to its target, and encounters least air resistance. Scrum-halves achieve this by holding the rugby ball in both hands and sweeping them past each other as they release the ball. The quarterback, in contrast, holds the more slender American football in one hand and throws it overarm, like a javelin thrower, sweeping the fingers across the ball as they release it, to impart the spin. Kickers in both sports often use a similar technique to control the movement of the ball when they kick from hand. They slant the long axis of the ball slightly obliquely so that when they kick the ball, it spins along its long axis and curves gently towards the side in which it is pointing, allowing the ball to move into touch farther up the field.

Many other sports involve players hitting a ball with a club, bat, stick, or racket. The way these hitting movements are powered is plainly very similar to the ways in which workmen power their hammers and soldiers swing their swords, and which we examined in chapter 15; they all make use of multiple sling actions. Take the most straightforward, for instance, the mechanics of the golf swing. More must have been written about the perfect golf swing than almost any other subject, but in essence it is actually incredibly simple, and the golfer has the great advantage that the ball is stationary. During the swing, the golfer first initiates motion by rotating their hips and their body, which causes their shoulders to rotate. This rotation accelerates the golfer's straight left arm with a slingshot action and causes it to swing around in a more or less circular arc. In turn, the club is then accelerated by a further slingshot action. By the time the golfer's left arm is pointing downward, the club is parallel to it and travelling at maximum speed when it hits the ball, after which it is slowed as the clubhead swings inward during the follow-through. Or at least that is the theory. The difficulty is in the timing. At the start of the golf swing the golfer has to hold their wrists rigid, to accelerate the club along with the arms. Only when the arms have got up to near full speed will the slingshot action draw the clubhead forward and power its acceleration. The golfer has to time exactly when to "release" their wrists, and decide how much additional power to apply to accelerate the clubhead further by actively rolling their wrists. Computer simulations show that

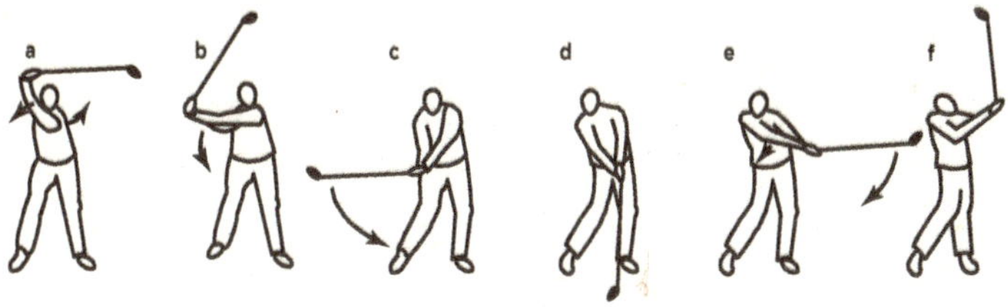

The mechanics of the golf swing. At the start, the wrists are held firm and the rotation of the shoulders (a) powers a sling action that drives the swing of the arms (b). Release of the wrists allows the arms to rapidly accelerate the golf club due to a second sling action (c) until contact (d), after which a reverse sling action slows the clubhead (e) and the swing ends in a suitable pose (f).

90 percent of the clubhead's speed comes from the slingshot action; at most 10 percent comes from the wrists. But the wrists also affect how square-on the clubhead is when it hits the ball. The looser the player's wrists, the more delayed the clubhead's rotation could be, resulting in a slice; the more powerful the wrists' action, the more the clubhead could be accelerated, resulting in a hook. Both are likely to end in disaster since the sideways spin imparted to the ball will cause it to curve away into the rough or out-of-bounds.

Racket sports demand an even greater control of spin. Tennis, squash, and badminton players all use the slingshot action to accelerate their rackets so that they are travelling rapidly at contact, just like a person wielding a hammer or a sword. And like swordsmen, racket players also aim to hit their projectiles at the "sweet spot" in the centre of the racket head, the equivalent of the sword's point of percussion, so that they experience no recoil at their wrists when they make contact. They produce both forehand and backhand shots by using a triple slingshot action, rotating their hips, body, and shoulders, which in turn swings the arms, and that finally rotates the racket head. In the tennis serve, the action is similar but more like the overarm throw of a javelin thrower. But once again, the real skill of an elite tennis player is in disguising the direction in which they are going to hit the shot. They can change the amount of power they apply to the racket to some extent by actively flexing their wrists, but a simpler method is to alter the motion of the hands and arms. Pulling the racket arm in a sharper curve inward as it swings forward accelerates the racket head faster, so it tends to hit the ball when it is farther forward relative to the body. It will produce a faster crosscourt shot. A good example of this sort of shot is the "whipped" forehand. In contrast, allowing the arm to move in a straighter direction will slow the acceleration of the racket head, which consequently hits the ball farther back relative to the body, resulting in a shot that is pushed more slowly "down the line." A good example of this sort of shot is a "sliced" backhand. The racket head can also be held at different angles to impart topspin or backspin to the ball, causing it to curve downward in its flight or travel in a flatter trajectory.

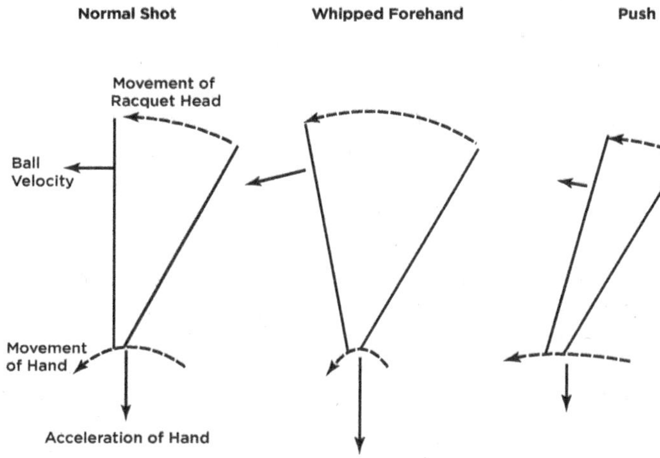

How altering the movements of the hand can change the speed and direction of tennis shots. In a whipped forehand (*centre*), the hand is accelerated inward faster, accelerating the racket to a greater speed so that it hits the ball harder and earlier. In a push (*right*), the inward acceleration is reduced, resulting in lower racket speed and hitting the ball later.

But of all ball games, the ones that demand the greatest mastery of rotation and the widest range of skills must be the sports of baseball and cricket. I'm no expert on baseball, having only played the British children's game of rounders, from which baseball derived, but even I realise that the feats of its pitchers and batters are prodigious. Batters swing their heavy bats two-handed (unlike the lighter single-handed bats used in rounders) using a slingshot action, rather like a sideways golf swing, and they aim to hit the ball at its sweet spot, around two-thirds of the way along the bat. I particularly admire the way the batters can propel a ball forward using such a narrow cylindrical bat; it must make it incredibly difficult to hit a pitch head-on. And the batter's job is made even more difficult by the speed and curve imparted to the ball by the pitcher. The pitch is essentially a straightforward throwing action, but taken to the extremes. Pitchers twist their bodies back and even raise a foot to allow extra hip rotation as they wind up for the delivery, so that when they unwind their bodies they rotate their shoulders at high speed. And to maximise the power of the sling actions, they often hyperextend their shoulders and elbows. Years of practice from an early age means that they develop less torsion in the humerus on the pitching arm than on their other arm. This allows them to produce a stronger slingshot action and accelerate the ball to speeds of over 105 miles per hour (almost 170 kilometres per hour). Moreover, pitchers can impart a range of spins to the ball by gripping it at different points of the figure-eight seam and rolling their fingers or wrists around it. In sinkers, they impart topspin to the ball; in rising fastballs they impart backspin; and in cutters they can give the ball spin to either side. I only wish I knew what was going on when I watch baseball games so that I could appreciate them more.

In cricket the complexities (and the bafflement of foreigners) are, if anything, even greater. The rules of cricket forbid bowlers from bending their arms when delivering the ball—"chuckers" are quickly rooted out and banned—so bowlers have reverted to a different method to bowl faster. They run up to the crease, leap into the air on their delivery stride, and rotate their shoulders, creating a single slingshot action that powers the rotation of the bowler's straight arm. A final flick of the wrist adds another small element to the slingshot action and means that fast bowlers can deliver cricket balls at over 90 miles per hour (145 kilometres per hour).

As well as bowling fast, the bowler makes it more difficult for batters to hit the ball to the boundary by bowling it so that it bounces around three-quarters of the way to the wicket. A delivery with "perfect line and length" will reach the batter a couple feet off the ground, high enough to make it difficult for the batter to hit the ball at the bat's sweet spot and encouraging them to hit the ball into the air, but low enough so that it could still hit the stumps. The only response a batter can make to this sort of delivery is a forward or backwards defensive shot, hitting the ball with a vertical bat that is angled downward so it hits the ball slowly into the ground. To score runs, the batter has to wait until the bowler delivers an inaccurate ball. If the ball bounces too close to the batter and is "overpitched," the batter can swing the bat freely forward in a vertical plane, catching the ball low on the bat at its sweet spot, and dispatch it forward to the boundary in a shot known as a "drive." On the other hand, if the ball bounces too far away from the batter, a "short" delivery, the batter can swing the bat horizontally in a "pull" or "hook" to hit the ball sideways to the boundary. And if the bowler delivers the ball too far away from the line of the stumps, the batter can hit the ball with their bat horizontal but slightly later so that the ball travels to the other side of the boundary in what is known as a "cut" shot.

In turn, bowlers try to deceive batters with their own trickery. One particularly devious delivery is the "yorker," in which the bowler delivers the ball so that it hits the ground next to the batter's feet, sneaks under the bat, and hits the stumps. If bowled accurately this delivery can be devastating; and even if the batter succeeds in hitting the ball it will be with the toe of the bat and will sting the batter's fingers. The only problem is that the yorker is high-risk. If it bounces a little too early it can be driven by the batter, and if it stays in the air it becomes a "full toss" and is even easier to hit. Bowlers, like baseball pitchers, also use a variety of techniques to change the direction in which the ball travels as it flies towards the batter. Seam bowlers can deliver the ball with the seam tilted slightly from the vertical so that it hits the ground at the edge of its seam and bounces slightly sideways; or they can shine one-half of the ball more than the other so that aerodynamic forces cause the ball to curve through the air. And spin bowlers rotate the ball with their fingers as they release it or allow it to roll over their wrists to make the ball spin even faster so that it bounces in unexpected directions. The range of spin deliveries is vast, from conventional "off breaks" and "leg breaks" to "top-spinners," "googlies," "flippers," and "doosras." All in all, this makes it possible for grown men to waste days of their lives watching their heroes battle it out in contests that might simply fizzle out into an inconclusive draw.

But the most amazing sportspeople of all are those who have developed ways of controlling the rotation not only of their arms and legs, or a range of clubs, rackets, and bats, but of their whole bodies: long jumpers, skaters, gymnasts, and divers. They seem even more than gyroscopes to defy the laws of physics and be able to alter the rotation of their bodies even without applying forces to the ground.

One of the best-known examples is seen in the feats of long jumpers. When a jumper takes off they push down hard with their takeoff foot, effectively vaulting over it to convert some of the horizontal speed of their run-up to vertical movement, giving them time to travel through the air before they land in the sand pit. The only problem is that the force the takeoff foot produces acts behind the centre of mass of the athlete's body so that it tends to rotate them facedown with the legs moving backwards, tending to make the jumper dive forward into the pit and reducing the length of the jump. One way some long jumpers minimise this effect is to raise their arms at takeoff in the "hang" technique, which increases the moment of inertia of their body and reduces the forward rotation. Another technique is the "hitch kick." After takeoff, the jumper continues to swing their legs as if they were still running, moving the legs back when they are fully extended, and moving them forward while they are bent. They can also do the same with their arms. The net effect of these movements is a facedown rotation of the legs and arms that reduces the rotation in the rest of the body, which remains upright until just before landing. At this point the long jumper bends their body, drawing both the upper body and legs forward so they land farther away from the takeoff board.

The hitch kick technique. To prevent the body rotating nose-forward, the jumper rotates their arms and moves their legs like a cyclist to produce net forward rotation of the limbs. Just before landing, they bring both legs and body together so that the feet land as far forward as possible.

Another method of controlling the speed at which you spin is to alter the moment of inertia of your body, a technique that is commonly used by ice-skaters as they come to the conclusion of their routines. They circle inward before settling into a spin, with their body and one leg horizontal, before gradually pulling themselves upright, bringing their legs together and finally raising their arms aloft, which has the effect of accelerating their rotation until, at the climax, they suddenly come to a dramatic stop. This process is invariably explained by invoking the concept of conservation of angular momentum. Textbooks explain that as the skater pulls her arms inward (it's always a "she"!), she reduces her moment of inertia. Because angular momentum is conserved, her whole body rotates faster. But this is just a hand-waving explanation. To understand what is really going on you have to think about the movements and forces involved. Take a simplified case in which the skater's hands are modelled by weights at the end of a string. If the skater pulls the weights inward, they will immediately move forward relative to their base for two reasons. First, the inward force will increase the velocity of the weights, which will now move in an inward spiral. Second, even if the velocity was the same, the weights, being closer to the centre of rotation, would have a greater

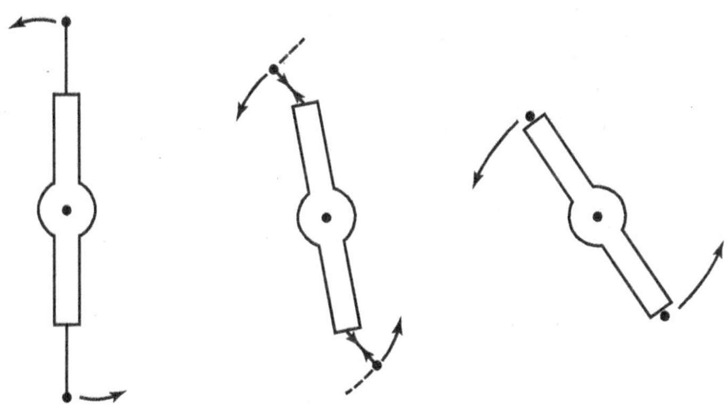

Why skaters spin faster as they pull their arms in. The hands move in an inward spiral and the centrifugal force on them drags the rest of the body around faster.

angular velocity. The string, tensed by the centrifugal force on the inwardly spiralling weights, will pull the inner part of the body forward. The same will happen with the real skater, whose arms will be slightly bent by the forces. And it's not just her angular velocity that increases; her whole body accelerates. The skater supplies the energy to do so from her muscles as she pulls her arms inward against the centrifugal force.

Gymnasts do much the same thing to help them swing faster when they are exercising on the horizontal or parallel bars. To start swinging in the first place, when they are hanging by their hands below a bar, gymnasts flex their bodies, which acts to accelerate them, just as bending our hips accelerates us when we are standing up. To increase the swing, the gymnast continues to swing their legs backwards and forward in rhythm with the natural period of oscillation of their bodies. And once they have started to swing, they can increase the amplitude of the swing by using the same mechanism that skaters use to rotate faster. They raise the centre of mass of their bodies and move them closer to the centre of rotation by raising the knees as they swing forward, producing an additional rearward force on the bar and pushing the body forward and upward. It's something that our ancestors probably did when they were living in the forest canopy, so it's no surprise that the process is most spectacularly displayed by the doyens of swing, those most delightful of apes, the gibbons. Native to the rainforests of Southeast Asia, gibbons move at speeds of up to 35 miles per hour (56 kilometres per hour) through the forest canopy using the mode of locomotion known as brachiation; they swing from branch to branch, flying through the air between swings and using their long slender hands merely as hooks. They supply the power using the muscles of their shoulders and abdomen, which raise the body and legs through each swing so that it speeds up as it rises towards the next handhold.

Divers and gymnasts can also change how fast they rotate by flexing their bodies even when they have nothing to hang on to, when they are flying freely through the air. If they start their flight with their bodies held straight, and rotating slowly along their long axis, about a line across their hips, they can speed up their rotation. The simplest way of doing this is to fold the body into the "tuck" position, which greatly reduces its moment of inertia and greatly speeds up the roll. Using this technique, an elite performer can achieve double, triple, or even quadruple rolls before extending their body out again to slow it down, ready for landing.

And they can perform an even more spectacular manoeuvre, which at first sight seems to be totally baffling. They can convert some of their body's rotation about their hips, into spin around the long axis, to produce what is known as "twist." To do this, the performer moves their arms asymmetrically, raising one while lowering the other. This would tend to rotate the body about its other axis, causing it to tilt, but this process is resisted by the angular momentum of the body. Instead, it precesses, just like a gyroscope. However, unlike gyroscopes, our bodies are so slender that the moment of inertia around our long axis from head to toe is far less than that around our waist. Consequently, the performer precesses far faster than the original rotation. A good performer can spin around several times before reversing the position of their arms to stop the motion before landing.

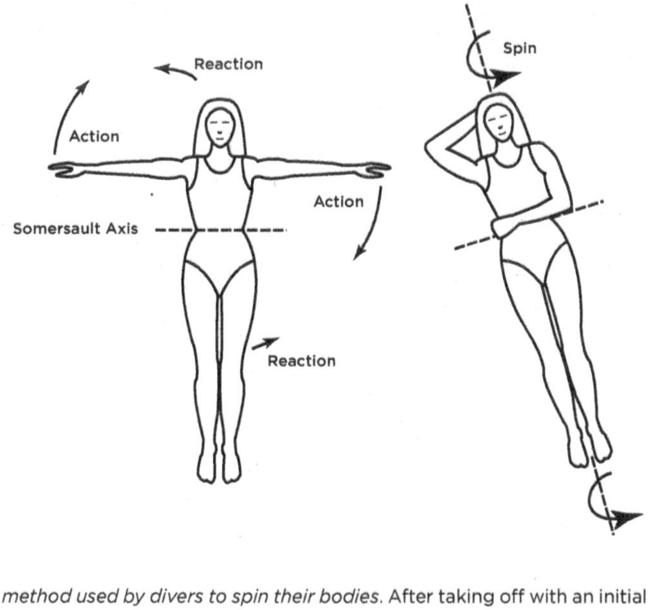

The method used by divers to spin their bodies. After taking off with an initial slow rotation about the somersault axis, the diver moves her arms asymmetrically to tilt the long axis of the body (*right*), causing it to precess about its long axis.

And remarkably, it is also possible to rotate the body even if it is not rotating when it starts its flight, a fact that seems to break the law of the conservation of angular momentum. The only problem is that you won't see this apparently law-defying manoeuvre in the Olympics, as the real experts are not people, but domestic cats. If you hold a cat with its body oriented horizontally and upside down and drop it (I don't personally condone this), it always manages to land upright on its feet, so in under half a second it has rotated its body by 180 degrees. It's a remarkable achievement, and given the popularity of cats online, it should be no surprise that falling cats are, like felines doing all sorts of other things, all over the internet. Some videos even set the movements of the cats to music.

Having seen how divers rotate their bodies around their long axis using precession, it is fortunately fairly straightforward to explain how cats perform this apparently impossible manoeuvre. The key is that cats are capable of flexing and extending their backs so that the two halves almost seem to act as separate elements. The cat arches its body upward during the first half of its fall so that each half rotates about the waist, but in opposite directions. If, during this process, the cat moves each set of legs asymmetrically—one forward, the other back—and the fore and hind legs move in opposite ways, as if it were trotting, these movements will tend to tilt each body half about the other axis, just as moving the arms asymmetrically tends to tilt the whole body of the human diver. The cat can also increase the force tending to tilt the body halves, by activating muscles on one side of its body, tending to bend it sideways. Just as in the whole body of the diver, the tilt of each half of the cat's body will be resisted by its angular momentum. Consequently, both halves of the body will start to precess; and since both the body rotation and tilt are in opposite directions, the two halves of the body will precess in the *same* direction! After a short while, the cat's body will have spun around by 90 degrees. In the second half of the fall, as the cat flexes its back downward again, it can reverse the movements of its legs and the sideways force on its body. Since the body halves and the legs are now both moving in the opposite direction from the first stage, the result will be to carry on spinning the cat, once again in the *same* direction. By the time it hits the ground, it will have spun around the full 180 degrees it needs to land upright; it will fall on its feet.

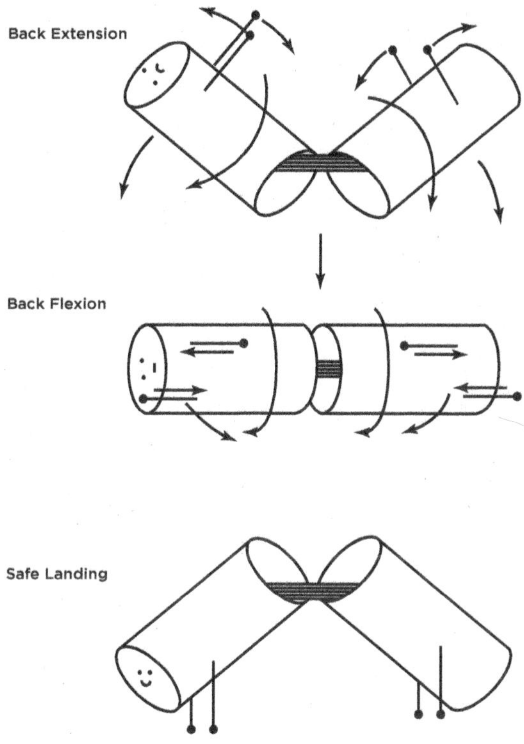

The rotation of a falling cat. At first the cat extends its back while moving its limbs asymmetrically, which causes both body halves to precess in the same direction. It then flexes its body while reversing the leg movements, which causes the body to keep precessing in the same direction. By the time it nears the floor it has rotated by 180 degrees.

It seems that using their multi-jointed bodies, humans and other animals have an almost inexhaustible set of techniques that they can use to control their motion. These far exceed the limited range of ways that we have managed to exploit in our own technology that is based on wheels and gears rotating about a single axis. And it is likely that the movements of other animals may exploit an even wider range of rotational tricks to help them move. Whether it be the galloping of horses, the flapping of birds' and insects' wings, or the swimming movements of fish, we still have a lot to learn about how organisms get about.

PART IV

PUTTING SPIN IN PERSPECTIVE

CHAPTER 19

The Chequered History of Spin

By rights, little of what I have written in this book should be new to a reasonably educated person, for the science of spin had, for all intents and purposes, been solved more than two hundred years ago. Indeed it was one of the first topics to be investigated by early scientists, and for a good reason. To provide support for the new Copernican model of the solar system, in which the earth spins around as it orbits the sun, they needed to explain how the system moved and kept going. And it all started so promisingly. The key conceptual breakthroughs were made by the great English polymath Robert Hooke in the year of the Great Fire of London, 1666. He was the first person to articulate the idea that a body moving in a circle travels at a steady speed while being accelerated at right angles to its motion. And he was also able to show, by performing a simple public demonstration, how a universal gravity could supply the inward centripetal force needed to keep the earth orbiting around the sun and the moon orbiting around the earth. He modelled the earth by using a ball that hung by a long string from the ceiling. To model its orbit, Hooke simply pulled the ball away from its hanging point and set it swinging in a circular motion. In turn, the moon was represented by a smaller ball hanging from the earth by a shorter string. This, too, he set in motion so that the moon could be made to rotate about the earth in the same direction as the earth moved around the sun, only faster. Hooke also went on to show how the centrifugal force on the earth due to its spin would deform it into a flattened ellipsoid that expanded outward at the equator. He modelled the earth again, but this time using a molten glass vessel and showed how spinning it around would cause it to bulge outward. And by analogy with the strength of light emanating from a candle, he even went on to suggest that gravity exerted by a body should fall with the square of the distance away from it: the famous inverse square law.

But analogy and demonstrations have their limitations. The machinery of the universe was put into a solid mathematical framework by the genius of Hooke's great rival and nemesis, the mathematician and mystic Isaac Newton. Basing his work on Hooke's conceptual framework, and using novel geometrical arguments, Newton was able to prove that a universal gravity that fell with the square of the distance from an object would cause the planets to move around the sun in just the way that astronomers had observed; they would follow Kepler's three laws of planetary motion, sweeping out equal areas of their elliptical orbit in equal time, and travelling more slowly the farther away they were from the sun. Moreover, Newton was also able to use mathematical arguments to explain the tides, quantify the bulge of the earth, and even predict how fast the earth should precess. In his universally praised (though almost unreadable and unread) book, the *Principia*, he set out the mathematical basis for the whole of astronomy and mechanics. Modern physics was born.

Over the next century, mathematicians on the continent of Europe, most notably the Swiss Leonhard Euler and the Frenchman Jean Le Rond d'Alembert, set out to translate Newton's proofs into conventional calculus, and to extend his mechanical analysis to include laws for the rotation of solids as well as for the translational motion of particles. Their work was summarised by the Frenchman Pierre-Simon Laplace in his far more readable work, *Celestial Mechanics*. This book also included proofs for the stability of rotating objects and equations describing the effect of the earth's rotation on the movement of water over its surface—the forces that Coriolis later showed were examples of the forces that are exerted on all objects moving around a rotating reference frame. By the end of the eighteenth century, therefore, all aspects of the behaviour of rotating bodies had been encapsulated into a small number of equations that described the way in which they moved and the effects of their movement. It was a triumph that demonstrated that mathematics is an essential component of physics, and that, in Galileo's words, "science is written in the language of mathematics."

In the eighteenth century, under the influence of Isaac Newton, the contribution of experiments and the importance of applying science to everyday concerns was also increasingly downplayed in favour of the idea of the philosopher scientist, absorbed in the mathematics. As a consequence, today the attitude that physics can only be understood using mathematics is all-pervasive, as I found when being taught as an undergraduate in the 1980s. And because the mechanics of spin and rotation only involves simple mathematics, their place in the curriculum are often reduced to around ten minutes' lecture time and three or four equations!

There are many advantages to this mathematical approach. It is precise and elegant, and findings can be fitted together to produce a comprehensive framework of knowledge. The mathematics can also be repurposed and used again as new sciences develop. For instance, in the twentieth century the equations of spin were successfully applied in quantum theory to explain the behaviour of subatomic particles such as electrons and muons. Particles, it appears, have different quanta of spin, and though the rotation is not real (for a start, electrons are infinitely small, so have no moment of inertia), this quantum spin does affect their behaviour in ways that the mathematics of rotation can predict. It is why some materials are magnetic, for a start. And quantum spin has even been found to be practically useful, as large magnetic fields can alter the precession or wobble of hydrogen nuclei in organic tissue, which is the basis of NMR and MRI imaging.

But the hegemony of mathematics has also had extremely harmful effects, especially when it comes to understanding many of the important aspects of classical physics, including rotation and spin, which we encounter in our everyday lives. Mathematical arguments exclude the majority of the population who are unable or unwilling to grind their way through them, which is one reason why I have not included any equations in this book. And even for the mathematically literate, they do little to explain what is going on. Consequently, physicists have for centuries struggled to work through the consequences of the laws of rotational motion, or to articulate them to scientists from other disciplines. This has greatly obstructed the progress of science, especially our understanding of the workings of the universe and of our own bodies, which I have outlined in parts 1 and 3 of this book.

The first, and perhaps most glaring failure, was the long delay in articulating the consequences of the laws of rotational motion. When Newton proposed his laws of translational motion in the *Principia*, he was quick to realise that the momentum of the universe—the sum of all the masses multiplied by the velocities of all particles—is conserved. Conservation of momentum proved to be one of the main planks in understanding dynamics. In contrast, it was not until the middle of the nineteenth century, a hundred and fifty years after Newton's *Principia*, that the term *angular momentum* was first coined and popularised by the mechanical engineer William Rankine. Only then did physicists realise that, like momentum, angular momentum had to be conserved. So one hundred years after the nebular hypothesis of the formation of the solar system had been proposed, this revealed that it had a potentially fatal flaw. If the sun had formed from a cloud of swirling gas, it would have had to have lost most of its angular momentum to have contracted to its present small size. It was not until forty years ago that physicists came up with a plausible mechanism about how it did this, and not until the last few years that the effectiveness of this mechanism has finally been confirmed.

Nowadays, people are also all too ready to accept that the powerful gravity field around black holes can swallow anything that comes near them. In fact, since angular momentum is conserved, if a particle of matter got closer to the black hole, it would speed up and swing out again in an elliptical path. A black hole should therefore be unable to swallow a particle however strong its gravity. The clue to how black holes swallow stars, in fact, lies in tidal forces. In the extreme gravity field near the event horizon of a black hole, gravity will not only be immensely strong, but the gravity on the near side of the star will be far greater than on the far side. This will set up extreme tidal forces, which tend to stretch the star into a lemon shape. As a result, the near side of the star will tend to orbit faster and the far side more slowly than its centre; this will stretch the star into a long curved string, a process called spaghettification. Eventually the star will get so stretched that it fractures; the outermost part of the string will break off and be slung outward into space, taking much of the angular momentum away with it and allowing the black hole to pull the inner part of the star farther inward into its grip. The process can then carry on; the black hole spits out more and more of the star as it draws it farther inward. The only reason black holes can swallow up so much material is because they are such messy feeders!

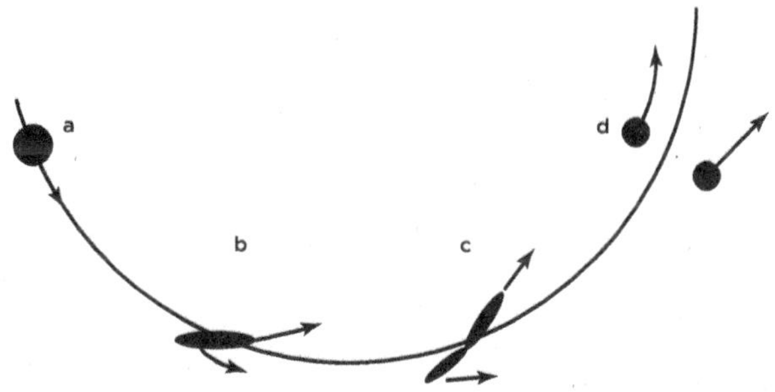

How black holes swallow planets and stars. Extreme tidal forces stretch the object until it eventually breaks, allowing the inner part to be pulled inward and slinging the outer part into space.

A second result of the obsession with mathematics is that it has hindered our understanding of how spinning objects behave. The precession of a gyroscope is traditionally explained by a single vector equation, which to most people is no help at all. This is probably the reason why Eric Laithwaite made his outrageous claims about gyroscopes breaking the laws of physics back in the 1970s. Fortunately, the great American physicist and hero of the Challenger inquiry, Richard Feynman, did produce the sensible explanation for precession that I shared in chapter 2. However, it is clear from reading his famous lectures that, having started from the equation, even Feynman struggled to turn his mind to the movements of a real-life object. And when he came to explaining why plates wobble as they spin through the air—as he did in his autobiography—he tackled the problem in a wholly mathematical way. This probably led to him mistakenly stating that plates wobble half as fast as they spin, whereas in fact they wobble twice as fast. This reluctance to think physically later led him to suggest that there was no intuitive physical way to explain why Dzhanibekov's wing nuts flipped, though as we shall see later in this book, this is in fact straightforward to understand.

Scientists' similar reluctance to think up convincing physical explanations for the behaviour of fluids moving in a rotating plane—how Coriolis forces are generated—have been even more damaging to science. Ever since 1735, people have been repeating a bogus explanation for the east-west motion of the trade winds, put forward by an English lawyer, George Hadley. He suggested that winds in the north-south direction veer to the side because of the different speeds at which the surface of the earth rotates at different latitudes. Not only is this explanation quantitatively wrong—it only explains half of the observed deflection—it fails to explain how winds veer off from the east-west direction, and hence fails to explain how circular weather systems form. Nevertheless, it has been pounced on eagerly by scientists reluctant to think about the physical reality. And intriguingly, this even included the mathematical genius Laplace, who had derived the equations that described how water currents are deflected on the earth's surface—and, which we saw in chapter 5, also govern the motion of the atmosphere. Even today, Hadley's incorrect explanation dominates scientific discourse; it is that given in Wikipedia, for instance, and I even recently heard it given on BBC Radio by the chief executive of the Royal Meteorological Society, who should surely have known better.

In the event, the great American pioneer of meteorology, William Ferrel, was finally enabled to identify the true cause of global wind patterns and rotating weather systems in 1854 thanks to a recent demonstration of Coriolis forces that had captured the world's imagination: Foucault's pendulum. The French physicist Léon Foucault was clearly fascinated by spin, as he was the scientist who gave the gyroscope its name, but his route to world fame was the huge wire pendulum he erected in 1851, first in the Paris Observatory, and then in the nearby Panthéon. Once released, the pendulum bob swung back and forth, as you would expect, but over time something odd happened. Rather than moving in the same back-and-forth direction, the plane of swing gradually rotated in a clockwise direction, as a trail of sand that dribbled through a hole in the bottom of the bob recorded; the pendulum bob only returned to its original orientation every 31.8 hours. The experiment was quickly hailed as proof of the earth's rotation. The pendulums drew massive crowds and soon Foucault pendulums were set up around the world. It rapidly became apparent that the time of rotation depended on the latitude: it was twenty-eight hours in Helsinki, which lies at a latitude of 60 degrees north, and forty-eight hours in Casablanca, which lies at 33 degrees north, while at the equator the pendulum underwent no deflection at all. The Foucault pendulum rotates because of the Coriolis forces on the pendulum bob, which in the Northern Hemisphere deflect moving objects to the right. As it is released and swings one way, it will be deflected a very small distance to the right until it comes to a halt at the far extent of its initial swing. The force of gravity will then draw it back again and it will once again veer to the right. When it returns to the original side it will have deflected slightly clockwise, and over time the track of the pendulum will gradually rotate, like a Spirograph drawing.

Ferrel realised that air currents would be deflected in the same way as the pendulum, and derived his revolutionary explanations that finally gave meteorology its firm scientific foundation. It is sad to report, though, that true to form, most modern descriptions of Foucault's pendulum fail to describe how Coriolis forces actually deflect

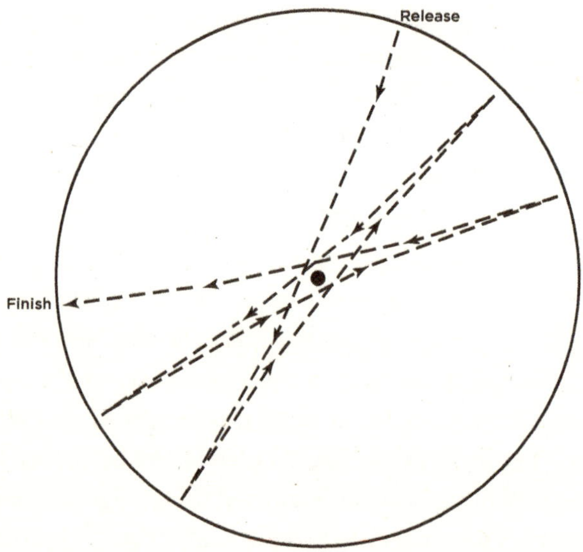

Movements of a Foucault pendulum bob, with the curvature of the path exaggerated for effect. Whichever way the pendulum travels it always swerves to the right, passing by the point where it would naturally hang (*solid circle*), and the path rotates.

its motion. Instead, they cling to the hand-waving argument, that the pendulum retains its angular momentum and swings in the same orientation while the earth rotates below it. This simply cannot be true; because the Paris pendulum took 31.8 hours to return to its original orientation, after a day it would only have deflected around 270 degrees, and would be oriented at right angles to its original orientation; its angular momentum would be totally different!

The most bizarre and troubling example of physicists' faith that mathematics can explain the effects of spin relates to an attempt to identify the cause of the earth's magnetic field. In the late 1940s, when Walter Elsasser in America and Edward Crisp Bullard in Britain were building our understanding of how Coriolis forces deflect the convection currents in the earth's core to produce an electromagnet, the British Nobel laureate Patrick Blackett was dreaming up an almost occult alternative. Excited by finding an apparent relationship between the angular momentum and magnetic field strength in the earth, sun, and a few stars, he suggested that a magnetic field was a fundamental property of a spinning body. He spirited up an equation and, with much fanfare, published it in the prestigious journal *Nature*. Being a member of the scientific establishment, he also managed to obtain scarce funding to test his theory. Needless to say, his experiments, which measured the magnetic field around blocks of heavy metals (including gold "borrowed" from the Bank of England), came up with a resounding negative and he was forced into a humiliating retraction of his ideas.

But the biggest negative impact of scientists' lack of intuitive feel for the science of spin has been in the fields of human biomechanics and sports science. As we saw in the third part of this book, human movements involve rotation of several joints at the same time. They cannot be described using a single equation and so have largely been ignored by physicists. In my physics training, I can remember being given just one example of applying mathematics to such complex movements: using matrices to work out the resonant frequencies of a compound pendulum. In more recent years, physicists have sought to perform computer simulations to investigate the actions of trebuchets and the movements of the golf swing, but the numerical solutions they generate are no substitute for thinking about what is actually going on. I have been unable to find comprehensible explanations for them in the scientific literature and have therefore had to work out for myself how we actually move: how bending our bodies can help us balance and accelerate; how the enhanced pendulum effect swings our arms and legs when we walk, run, and kick balls; how sling action sequentially accelerates our upper arms and forearms when we do a karate chop and throw spears; and how hammers, swords, and ropes can magnify these actions by providing an extra stage to the sling action. I hope that in the future, biomechanics will start to use these concepts in their research and when they teach their students about how we move. They could finally replace the confused concepts that litter their papers and textbooks: concepts such as "kinematic chains" and the "sequential interaction principle." They could realise that when a joint is accelerating it can power the rotation of elements distal to it, even without the action of muscles.

Fortunately, this confusion has not affected how we move about in normal life, but there is one example of the damage scientists can do, from the Soviet Union of the 1920s, when Russian researchers were seeking to improve the efficiency of their industry. The ideology of the Bolsheviks was, of course, bound up with the ideal of the virtues of noble workers and peasants, hence the depiction of the hammer and sickle on the Soviet flag. So when in 1920 the Central Institute of Labour (CIT) was founded under the patronage of Lenin himself, hammering was one of its main focuses. The avowed aim of the institute was to develop a scientific approach to work management. In practice, under the guidance of its founder, the revolutionary, former metal worker and avant-garde poet Aleksei Gastev, this meant trying to train workers to act like machines. The identically dressed trainees would be marched to machines and set to work in response to buzzers. They were trained to hammer in the "correct" way by holding a hammer attached to and moved by a hammering machine so that after half an hour they had internalised its mechanical rhythm. Gastev believed that machines were superior to people, so this process would help to improve humanity. Unfortunately, the action of the hammering machines followed Gastev's romantic ideal of the perfect technique. In his view, the hands and arms of the worker should be moved "efficiently" parallel to the hammer, and all of them should move in a single plane directly in front of the body, as was commonly depicted in Soviet propaganda posters.

One of the few proper scientists at the institute, the biomechanic Nikolai Bernstein, recorded the movements of Gastev's hammering technique, using a camera and stroboscope to produce a "cyclogram," a series of superimposed pictures that record the movement over time. He compared them with the movements workmen naturally adopted when they were asked to undertake a hammering task; they swung their arms out much more to the side and rotated their elbows, before finally straightening them as they brought the hammerhead down. Bernstein found that the "correct scientific" method was far worse than the natural technique; despite moving their arms and hands faster, the workers trained at CIT produced a speed of the hammerhead 20 percent lower than workers who swung their hammers in a natural way, which, as the cyclograms show, exploited the usual benefits of a triple sling action. Unfortunately, under Soviet rule, when science disagreed with ideology, it was science that had to make way. Bernstein was forced to resign from the institute and workers continued to be taught to hammer in the new unnatural way. Fortunately, Bernstein's story has a happy ending; he went on to work in research well into the 1950s. Gastev was not so lucky; he was arrested in Stalin's Great Purge and was shot in April 1939.

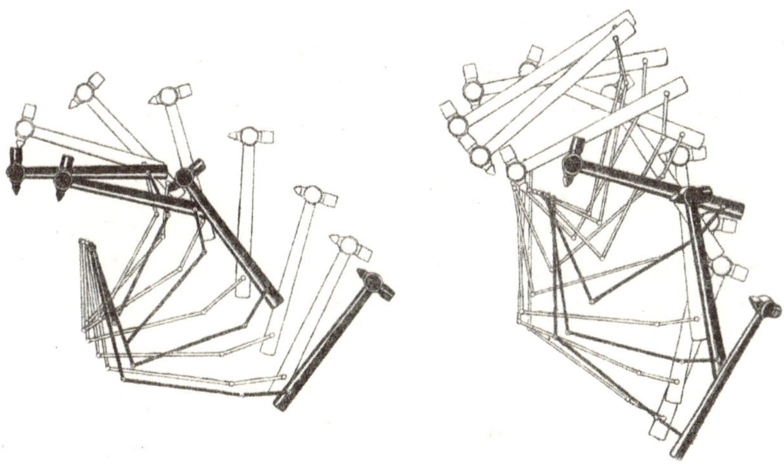

Bernstein's cyclogram comparing the movements of the arm and hammer during normal hammering (left) *and the "correct" technique advocated by Gastev* (right). Light symbols show the upswing and dark ones the downswing. Note the steady, rapid acceleration of the hammer using the natural technique, powered by a multiple sling action. In the "correct" technique the wrist remains too far cocked for much sling action to occur.

The biomechanics and sports scientists of today have not damaged modern industry to the same extent, but it is worrying to see how little progress they have made in understanding human movement. Despite using sophisticated camera equipment and computer software and spending hundreds of millions of dollars of research funding, modern researchers seem to have made little conceptual progress since Bernstein's day. And just as back in Bernstein's day, the workers knew that Gastev's ideas were flawed, so today sports biomechanics seems to be largely ignored by sports coaches, who have an instinctive feel for the best techniques. I remember my mother, for instance, a PE teacher and tennis coach, showing us how the power of a tennis shot came all the way from the floor and up through the body.

If the science of rotation and spin are misrepresented and misunderstood by researchers, educators simply ignore the technologies that use spin. The workings of reciprocating machines such as steam engines and internal combustion engines are widely described in children's books and in industrial museums. In contrast, there is seldom any mention of the technology that we rely on to run our modern world: the water, steam, and gas turbines, electric motors and generators, centrifugal and axial pumps and compressors; the milling machines, ring spinners, looms, and rollers. I have to confess that I knew virtually nothing about many of these vital machines until I started to collect information to research the writing of this book; and I have never met anyone other than a mechanical engineer who knew what a Francis turbine was or how a centrifugal pump works. It seems that despite the legacy of four hundred years of science, billions of dollars spent on research, and being subjected to over a decade of formal education, the average person is left just as much in the dark about how the world of spin actually works as our ancestors. Even though we live in a technological age, using a vast range of rotating devices, we know next to nothing about how they actually work.

CHAPTER 20

Spinning a New Yarn

Bearing in mind the confusion I outlined in the last chapter, it is not surprising that the word *spin* has been co-opted in recent decades by political journalists to describe the panoply of obfuscation, misrepresentation, and plain lies told by the press officers and PR gurus of politicians: the dreaded spin doctors. But fortunately, for science at least, there is no need to despair. For once we can put our finger on the reasons why people have found spin so confusing, we can also come up with a way to help us become better informed.

As we have seen, one of the main reasons for our ignorance is that because physicists rely so heavily on equations, they have failed to think hard enough about spin or to consider the movements of each part of rotating objects. This was plainly true of Eric Laithwaite as he gave those Christmas lectures back in the 1970s and claimed that gyroscopes broke the laws of physics. So I am happy to report that following the furore he created, he retired back to the laboratory and took a really close look at what gyroscopes are actually capable of. Working for years with the physicist William Dawson, in 1999 he came up with a patent that consisted of a box that held two gyroscopes that could be swung back and forth. The patent described how, simply by moving the gyroscopes in one particular order, the box could be made to move sideways. It seems that they had found a way to create a permanent linear displacement of an object without needing to apply external forces to it, just as falling cats have found a way to create a permanent rotational displacement of their bodies. Both do something that appears to be impossible, while still actually obeying the laws of physics.

Probably the best way to allow schoolchildren and university students to understand spin would be to allow them to take a more hands-on approach. We could give students the opportunity of actually playing with gyroscopes, fidget spinners, Frisbees, spinning glasses of water, and the like. This might make physics lessons more unpredictable, but it would certainly make them more fun and set students thinking, rather than just showing them equations. They could examine the behaviour of everyday objects, and our own bodies in particular. Children could be set to investigate how we balance and set off in motion, by looking at the effects of bending our bodies at the hips in different directions and tilting balance bars. And once they had worked out how those movements work, they could be set to study how they power playground swings. As most children know, the way to power a swing is to rock your body back and forth. The mechanics of propelling a swing when seated are essentially identical to those we employ when we bend at the hips to accelerate our bodies forward and backwards from a standing start. The only difference is that the whole process is inverted. To propel themselves forward, children need to accelerate their bodies face-forward. To do this, they need to pull back with their hands on the rope or rod that suspends the swing. In contrast, to propel themselves back, children need to accelerate their body face-backwards by pushing forward on the ropes. If these movements are made alternately, they can be made to propel the swing; the propulsion they produce will be maximised when the body is leaning back the farthest as the swing is moving forward, and leaning forward farthest as the swing is moving backwards. This means the rotation of its body is delayed 90 degrees behind the rotation of the ropes.

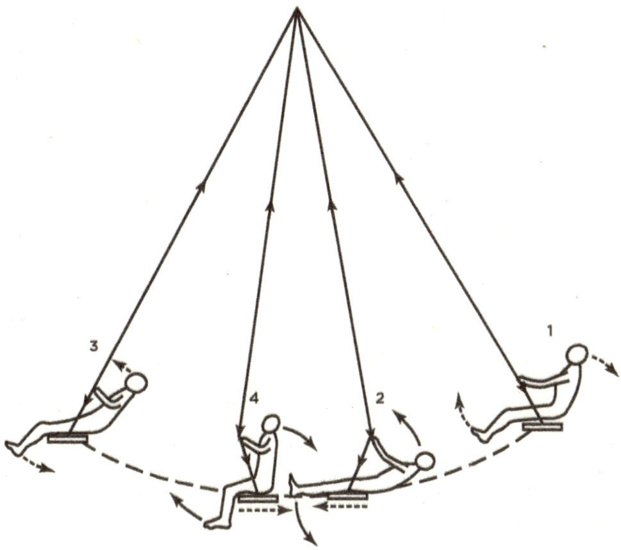

How a child propels a playground swing. At the back of the swing (1), the child is rotating backward (*dashed arrows*), but during the forward swing (2), she pulls back on the chain to accelerate nose-down (*solid arrows*), and the tension in the chain accelerates her forward. By the time she has reached the front of the swing (3), she is rotating forward (dashed arrows), before on the backswing she pushes forward on the chain to accelerate nose-up (*solid arrows*), and the tension in the chain accelerates her backward.

In order to understand the way we power the swings of our legs, children could investigate the enhanced pendulum effect by playing with the "clackers" or "ker-knockers," toys that became a craze back when I was a child in the 1970s and which have made short comebacks in the 1990s and (in Egypt) in 2017. Clackers consist of two plastic balls on the two ends of a piece of string. Holding the string at its centre, the user can move their hands up and down to make the balls act like enhanced pendulums. If the hands are moved up and down fast enough, the two balls can be powered so that they hit each other at the top of the swing as well as at the bottom, and they make a machine-gun-like rapid rattle as they hit each other. These toys therefore demonstrate the phenomenon of enhanced pendulum action in a particularly spectacular (and to parents, annoying) fashion. And they demonstrate very clearly that you don't have to apply torques at joints to get things rotating.

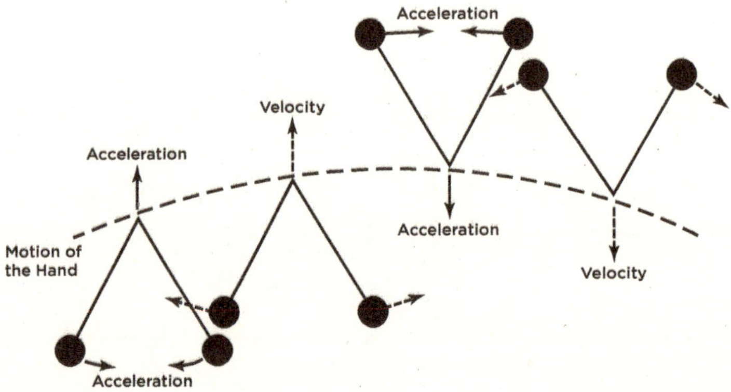

The action of the clacker toys familiar to those with 1970s childhoods. By accelerating your arm up and down, you can cause the balls to knock into each other both above and below your hand, creating an almighty racket.

In order to understand the way we power the swings of our arms when we hit or throw things, the best toy to investigate would be a dog ball thrower. Students could investigate how far they can throw a tennis ball with throwers of different lengths, and compare rigid throwers with flexible ones, to see how effectively they could store energy to add to the power of a throw.

But there is a problem even with taking this experimental approach. Many people find spin baffling because it tends to operate over nonhuman timescales. One reason why it took so long to understand the movements of the planets and stars is that they orbit and spin so slowly. It's hard to believe that movements we cannot see or feel can have such great effects on our lives. In contrast, the other instances of spin in our lives happen too fast. Most modern machinery rotates many times a second, so their moving parts are just a blur. And even when we move about, walking, running, and wielding tools, our actions typically occupy less than a second; they happen far too quickly for us to grasp what is going on in our bodies.

What we really need in order to help our understanding are ways of modeling situations and presenting them in easy-to-understand forms. The first, and best-known, model that sought to overcome some of these problems was the orrery. This clockwork model of the solar system was invented right back at the end of the seventeenth century, and an early model was presented to Charles Boyle, 4th Earl of Orrery, hence the name. Demonstrating the orbit of the known planets, the spin of the earth and the orbit of the moon, it showed admirably how the planets move, if not why.

And nowadays we are also fortunate to have the tools we can use to slow down the movements of fast-rotating objects. Never has it been easier to make slow-motion or time-lapse videos that convert the time course to a human scale: one that enables us to follow what is happening and for a commentary to give viewers a real feel for the mechanics involved. And never has it been easier and cheaper to distribute these videos by uploading them onto platforms such as YouTube. One admirable example is the explanation of the Dzhanibekov effect by S. Abbas Raza, which is available on the Veritasium website. As Raza points out, the Swiss mathematician Leonhard Euler had investigated this over two and a half centuries ago. Using ingenious mathematical arguments, he showed that for an object with three mutually perpendicular axes of rotation, stability depends on the moment of inertia about each axis. For an object such as a sphere, in which each moment of inertia is the same, the object will spin stably in all orientations. But if the moments of inertia are different, the object will spin stably only around the axes with the greatest and the least moments of inertia. Spin around the intermediate axis will be unstable.

You can see this for yourself by spinning a cuboidal object like a mobile phone, an elliptical fishplate, or even a more irregular object such as a tennis racket. The phone will spin stably when spun around its centre line down through the screen or when spun around its short axis. But if you try to spin it around its long axis, it refuses to do so and starts tumbling.

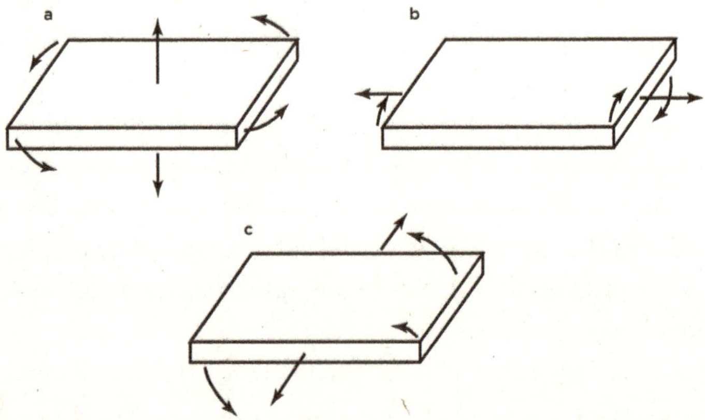

Objects such as mobile phones spin stably around the axes with the greatest (a) and lowest (b) moments of inertia, but not around the intermediate axis (c).

The reason why your phone tumbles, and why Dzhanibekov's wing nut flips is clearly shown in Raza's video, which animates the explanation put forward by the mathematician Terry Tao. If the rotating phone or wing nut tilts, the centrifugal force on the two sides will set up a moment that tends to *increase* the tilt and it will tend to swing 180 degrees over before rotating in its new unstable orientation for some time. It will keep flipping over, apparently randomly, until finally it settles into a stable orientation and spins through the axis perpendicular through its centre.

Educationalists are also starting to produce a wide range of videos that explain similar aspects of rotation: why gyroscopes precess and why spinning plates wobble, for example. They could also film the experiments that demonstrate Coriolis forces: how spinning a glass of water affects how it behaves. All you would have to do would be to film the proceedings with a camera that rotates with the glass. The same effect can be demonstrated by examining the movements of a ball rolling on top of a spinning parabolic mirror. Meanwhile, the effects of Coriolis forces on the earth's weather and climate could be demonstrated by showing time-lapse videos of the earth as seen from space; you can readily follow rotating weather systems such as depressions and hurricanes as they develop. Simulations of how the tides sweep across the face of the globe would be just as illuminating.

Grady Hillhouse on the Practical Engineering website has also made an excellent video showing how yo-yo de-spin slows down the rotation of satellites. We need more videos that show the action of devices such as trebuchets that use similar technology.

But as well as understanding the mechanics of spin, we also need to publicise just how crucial a role spin plays in our everyday lives, keeping our modern society rolling along. A second set of videos could illustrate the workings of rotating machinery. They could use cutaways to see the movements of the fluids and the propellers and impellers in turbines and pumps, and the action of differential gears and universal joints in cars. They could also show real footage showing the actions of rollers in shaping steel plates and girders, and the movements of textile machinery.

Enlightening films could be made showing the sheer number of rotating devices we use. In our homes alone there are the centrifugal pumps that drive water around our washing machines and dishwashers and out of the heads of our power showers; the ones that pump hot water through our radiators and refrigerants around our fridges and air conditioners; the ones that pump air to the burners of our boilers, through hair dryers, and into the cylinders of vortex vacuum cleaners. There are the electric motors that power all these devices as well as fan ovens, coffee grinders, and food processors in the kitchen; hard drives in our computers; lawn mowers, edge trimmers, and hedge trimmers in the garden. Our conventional automobiles are started by small electric motors and fast-spinning turbochargers pump air into their cylinders, while modern electric cars are powered by large electric motors that double up as dynamos to store energy as the car brakes. A film could also demonstrate how crucial rotating devices are for our infrastructure: the turbines and generators that produce our electricity; the pumps that raise water and drive it to our homes and that pressurise gas and set it moving across continents. The pumps that mix the chemicals in our refineries and stir the ingredients in our food factories. The rolling mills that grind our grain. And the milling machinery, lathes, potter's wheels, and circular saws that shape so many of the machines and artifacts that we use every day. These films would show that without spinning devices, our day-to-day lives in our modern world would grind to a halt.

And finally, and most important of all, we need to make slow-motion videos of the movements of real people, demonstrating how the enhanced pendulum effect helps us walk, run, and kick, and how we use sling action to throw objects and wield tools. It could be the start of a whole new evidence-based science of biomechanics and sports science, one that would finally give us useful insights about how we can move more safely and how athletes can improve their performance. These techniques could also prove to be the basis for a whole new technology, using our bodies as models to develop machines that move about more efficiently, balance more effectively, and handle tools more carefully than the clumsy ones made from interlocking wheels and cogs that we have used for so long. We can do it all if we only we play about more with our hands and let our heads get into the spin.

Acknowledgments

This book is the product of my lifetime obsession with everything to do with spin, from spinning tops, bouncing balls, wheels, aircraft, and machinery, to sport, not of course forgetting cats. I am grateful to my former school, Hampton Grammar School (now Hampton School), for giving me a solid grounding in physics, through teaching the excellent Nuffield Physics course and the mechanical aspects of applied mathematics. Thanks to them I left school thinking I would become a physicist. But I must also thank the University of Cambridge Natural History Tripos for mathematicising physics so much that I had to make a diversion into zoology. It meant that I spent a happy career investigating the mechanics of the animals and plants I love, and examining the physics of the beautiful environment in which we live. It meant that I could remain in the magical worlds of nature and classical physics, about which we still have so much to learn.

I would like to thank the many students to whom I taught biomechanics at the Universities of Manchester and Hull, and in particular to Jane Mickelborough, who introduced me to human biomechanics and gait analysis, and to Hannah Taylor, who introduced me to the mechanics of throwing. In the writing of this book, thank you to my agent, Peter Tallack, for tightening up my synopsis, and especially to my editor, Colin Harrison, for reading earlier drafts and once again gently pointing out where I had lost my way down dreary backwaters. Also to Vivek Venkataraman for reading and commenting on part 3. Thanks, too, for the support of Emily Polson and the production team at Scribner for their efficiency and professionalism.

Once again, I must thank my partner, Yvonne, for her continued support and forbearance, even when I was playing around in an embarrassing way with sticks, travelling down information trails on my phone or computer, or disappearing into the attic to write.

I dedicate this book to the memory of my late parents. My mother, Louie Ennos, was a fine physical education teacher and tennis coach and an intuitive gardener, who passed on her love of movement and of the natural world. My father, Tony Ennos, was a creative physicist, and a pioneer of laser speckle interferometry. He taught me so much about mechanics, engineering, and DIY, and passed on his fascination with the world about us right up to his recent death at the ripe old age of ninety-six.

Notes

Most of the information set out in this book, particularly historical facts and statistics, is freely available online, through a wide range of websites. Some of the notes refer you to these sites. However, facts are just the building blocks of real knowledge and understanding. In some cases, therefore, I refer you to more useful information in the form of original research articles, reviews, and books that have aimed to link such information to tell stories about what we know, how we know it, and why we believe it is true.

PROLOGUE **CAMELS IN A SPIN**

xi *It was the first of many fatal crashes*: For the story of the Sopwith Camel see Robertson (1970).

xiv *the 1974 Royal Institution Christmas Lectures for young people*: See Laithwaite (1980).

xiv *a spinning wing nut flips its orientation by 180 degrees every few seconds*: See, for instance, this YouTube video: https://www.youtube.com/watch?v=1x5UiwEEvpQ.

CHAPTER 1 HOW SPIN CREATED THE WORLD

3 *In the Bible narrative, for instance*: Genesis 1.

4 *We might well agree with Voltaire's Dr. Pangloss*: See Voltaire's (1759) *Candide.*

6 *The rotational equivalent of mass*: The moment of inertia is actually the sum of the mass of each part of an object multiplied by the square of its distance from the centre of rotation.

8 *The grand tack hypothesis*: See Walsh et al. (2011).

10 *NASA's Parker Solar Probe*: See https://www.nasa.gov/feature/goddard/2019/nasas-parker-solar-probe-sheds-new-light-on-the-sun.

11 *As well as suggesting the nebular hypothesis, Immanuel Kant*: See the recent translation of Kant's short and surprisingly readable book, Kant (2012).

CHAPTER 2 HOW SPIN MADE THE EARTH HABITABLE

14 *it took the genius of Isaac Newton*: See a translation of Newton's *Principia*, Newton (1999).

CHAPTER 3 HOW SPIN STABILISES THE EARTH

22 *What initially happens when you release a spinning gyroscope*: For the full explanation, plus the mathematics, see Richard Feynman's lecture, Feynman (1964).

CHAPTER 4 HOW SPIN SHIELDS THE EARTH

29 *experiments that the British physicist G. I. Taylor performed*: See Taylor (1922). In fact, the phenomenon was first observed by Lord Kelvin in 1868, but not published. A good summary of these experiments can be read in Persson (2001c).

29 *Just like the sea lying on the ellipsoidal surface of the earth*: Away from the equator and poles, gravity on earth, which points directly to its centre, is angled slightly off vertical, causing a small force acting towards the poles. This exactly cancels out the centrifugal force on the water caused by the rotation of the earth. See Persson (2001b).

30 *behaviour that was first comprehensively analyzed in*: See Coriolis (1835).

33 *How it produces a magnetic field*: For an expanded account, see Carrigan and Gubbins (1979).

34 *It coincided with ecological disruption*: See Cooper et al. (2022).

CHAPTER 5 HOW SPIN CONTROLS THE EARTH'S CLIMATE AND WEATHER

39 *the spin of the earth does deflect the winds*: See Persson (2005).

39 *winds travelling east or west will only be deflected upward or downward*: This is known as the Eötvös effect. See Persson (2005). It was discovered by ships charting the variations of the strength of gravity across the globe. They found that gravity measurements were lower when ships were steaming eastward than when they were stationary, while they were higher when they were steaming westward.

39 *convection currents get diverted*: The breakthrough paper is Ferrel (1856).

43 *the Coriolis forces also explain the* local *weather patterns*: First described by Ferrel (1856) and nicely described in Persson (2001d).

CHAPTER 6 SPINNING AND DRILLING

50 *because the material of which they are made is split up into large numbers of isolated fibres*: See Ennos (2012a).

51 *The earliest twisted yarn found to date comes from the Dzudzuana Cave*: See Kvadze et al. (2009).

53 *The result was the invention of the first spinning tops*: For more information on tops of all sorts, see Gould (1975).

54 *The bow drill appeared as long as ten thousand years ago*: See Hodges (1970).

55 *A tomb in Mehrgarh, Pakistan, dated between 7000 and 5500 BC, contains teeth that have been drilled with small holes*: See Coppa et al. (2006).

56 *recent Neolithic finds from Israel*: See Goren-Inbar et al. (2012).

CHAPTER 7 THE UNDERWHELMING WHEEL

62 *in the New World the principle was never developed*: For further discussion of the factors influencing the development of wheels, see Ennos (2020).

CHAPTER 8 SHAPING UP

73 *the potter's wheel*: For more on its development, see Hodges (1970).

79 *the spinning wheel*: For more on the development of the spinning wheel, see Leadbeater (2008).

CHAPTER 9 BUILDING MACHINERY

83 *The first, and simplest, was the saqiyah*: More recently the same principle was used in the pumps of sailing ships. See Davies (2002). The sailors used a crank to lift water up from the bilge using a chain of close-fitting disks that pulled slugs of water up a pipe and overboard via a gutter on the deck.

83 *farmers came up with a whole range of new rotating devices*: More details about the machinery described in this chapter can be found in Hill (1984), Hodges (1970), and Smith (1975).

91 *The first solution to be devised was the cam*: For more on cams and cranks, see Farrell (2020).

93 *Al-Jazari came up with a series of beautiful geared devices*: See Hill (1984) for the best descriptions.

CHAPTER 10 THE INDUSTRIAL REVOLUTION

103 *To power machinery directly*: For more on Watt's rotative engine, see Uglow (2002).

CHAPTER 11 TURBINES, PUMPS, AND GENERATORS

112 *The young British engineer John Smeaton*: For more on Smeaton's work, see Smith (1981).

113 *The first practical benefits of this new understanding were supplied by a hydraulic engineer, Jean-Victor Poncelet*: For a more detailed account of the development of turbines, see Smith (1980 and 1981).

CHAPTER 12 GOING FOR A SPIN

137 *The real reason for the stability of bicycles*: See Jones (1970).

139 *Robert Hooke pictured a universal joint in his* Micrographia *of 1665*: See Hooke (2007) for a nice facsimile edition.

CHAPTER 14 STANDING AND STARTING

154 *our ancestors may well have acquired the ability to walk bipedally when they were still living in trees*: For a fuller up-to-date story see DeSilva (2021).

156 *the best way to avoid falling over*: See Winter (1995), which includes a description of the hip strategy, but no explanation of how it works.

159 *You might think it would be a simple matter to start walking*: For the conventional explanation see Winter (1995) and Mickelborough et al. (2003).

CHAPTER 15 **WALKING AND RUNNING**

164 *we can model our legs as two rigid inverted pendulums*: For the basic mechanics of walking, see Alexander (1982).

165 *Working in the late 1980s, McGeer made a simple 2D walker*: See McGeer (1990).

165 *Steven Collins of Cornell University combined aspects of the two models*: See Collins et al. (2001).

165 *He added a simple power control unit to his robots*: See Collins et al. (2005).

166 *his robots strolling along a corridor*: See, for instance: https://www.youtube.com/watch?v=v2qILGYl-BM.

167 *Steven Collins was also able to reproduce this motion in his robots*: See Collins et al. (2009).

169 *The normal arm swing also has benefits*: See Collins et al. (2009).

169 *our Australopith ancestors walked in the same way as we do*: See Crompton et al. (2011).

170 *It's a strategy the Dutch robot builder Martijn Wisse of Delft University of Technlogy used to power his own robot*: See Collins et al. (2005).

171 *At first glance, running looks very similar to walking*: For the basic mechanics of running, see Alexander (1982).

CHAPTER 16 **HITTING**

178 *The great apes all make a host of wooden tools*: See Ennos (2020) for more details.

178 *our closest relatives, the chimpanzees, use lumps of both wood and stone as simple clubs*: See Luncz et al. (2022).

179 *By around 4 million years ago, hominins such as* Australopithecus afarensis *had evolved longer thumbs*: See, for instance, Tocheri et al. (2008).

179 *well before the earliest stone tools were made, 3.3 million years ago*: The oldest finds were from the shores of Lake Turkana as described in Harmand et al. (2015).

179 *This would have allowed them . . . to use clubs to hunt and for intergroup warfare*: As suggested by Young (2003).

CHAPTER 17 THROWING

185 *Warwolf was the end result of almost two thousand years of technological development*: For a nice history of the trebuchet, see Chevedden et al. (1995).

186 *Hominins had probably been throwing missiles for millions of years*: Again, as suggested by Young (2003).

188 *the fastest a powerful chimpanzee can throw overarm is a mere 20 miles per hour*: See Roach et al. (2013).

188 *starting the action from the legs to power the first stage of the sling process*: The most influential recent paper, Roach et al. (2013), failed to identify the role of sling action, so instead suggested that elastic storage must help power the throw.

189 *It would have helped them fend off predators, hunt game, and fight rival tribes*: Again, as suggested by Young (2003).

189 *evidence, in the form of wear and wood residues on hand axes*: See Keeley and Toth (1981).

189 *The first undisputable hunting spears were those found in 1994 in Lower Saxony, Germany*: See Thieme (1997).

189 *experiments have shown that they can be thrown accurately distances of at least 70 feet (21 metres)*: See Milks et al. (2019).

193 *primitive-technology enthusiasts can throw spears up to 200 feet (60 metres) using a baton*: See Westcott (1999).

195 *over 75 percent of its energy is used*: For more on the technology behind the trebuchet, see Chevedden et al. (1995).

198 *NASA mathematicians even managed to analyze this process*: See Fedor (1961).

CHAPTER 18 THE SULTANS OF SPIN

200 *Why this is the case was demonstrated by a series of experiments*: See Ennos (2012b).

200 *elite athletes performed better than average ones in balancing on small seesaws*: See Ennos (2012b).

210 *Another technique is the "hitch kick"*: It's worth noting here that stunt drivers in films do exactly the opposite of long jumpers when they are asked to drive cars off ramps. Normally such a car would retain its orientation and fly nose-up until it landed on its rear wheels. Unfortunately, this does not look good on film. To make the stunt look more spectacular, the drivers slam on the brakes immediately after take-off. The angular momentum in

the spinning wheels is thereby transferred to the car, which rotates nose down and falls with a satisfying crunch onto its front wheels.

212 *those most delightful of apes, the gibbons*: See Fleagle (1974).

214 *the real experts are not people, but domestic cats*: See Gbur (2019) for much more detail about attempts to explain how cats turn over while they are falling, though amazingly he failed to identify the role of precession.

CHAPTER 19 **THE CHEQUERED HISTORY OF SPIN**

219 *The key conceptual breakthroughs were made by the great English polymath Robert Hooke*: For the history of Hooke and Newton's relationship and their relative contribution to gravity and the laws of motion, see Gribbin and Gribbin (2017).

220 *the* Principia, *he set out the mathematical basis*: For a modern translation and introduction to Newton's work, see Newton (1999).

220 *Laplace in his far more readable work,* Celestial Mechanics: Best read in the marvelous 1831 translation by Mary Somerville, Somerville (2012).

221 *the equations of spin were successfully applied in quantum theory*: See Tomonaga (1977) for a history of quantum spin, though I was unable to understand a single word of this book.

224 *it is clear from reading his famous lectures*: See Feynman (1964).

224 *when he came to explaining why plates wobble as they spin through the air*: See Leighton and Feynman (1992). In fact spinning plates wobble for the same reason that gyroscopes precess. When a plate tilts, the material at the rim that is moving at right angles to the axis of tilt moves in a curve, and reaction forces accelerate the plate in the opposite direction. The plate starts to tilt at right angles to the original direction, and the wobble rotates in waves around the rim.

224 *people have been repeating a bogus explanation for the east-west motion of the trade winds*: The confusion has been admirably charted by Anders Persson. For the most appropriate pieces of his work, see Persson (2005) and Persson (2006).

226 *The most bizarre and troubling example of physicists' faith*: For a fuller account of the Blackett affair, see Nye (1999).

227 *concepts such as "kinematic chains" and the "sequential interaction principle"*: For instance, in such popular textbooks as Knudson (2007) and Blazevich (2017).

228 *the damage scientists can do, from the Soviet Union of the 1920s*: For a good account of this affair, see Rose Whyman's recent translation of Bernstein's book, Bernstein (2020).

CHAPTER 20 **SPINNING A NEW YARN**

231 *in 1999 he came up with a patent*: See Laithwaite and Dawson (1999).

232 *children power their swings by rocking their bodies back and forth*: The mathematics, but not the explanation, are given by Case and Swanson (1990).

235 *which is available on the Veritasium website*: See https://www.youtube.com/watch?v=1VPfZ_XzisU.

237 *an excellent video showing how yo-yo de-spin slows down the rotation of satellites*: See https://www.youtube.com/watch?v=ZKAQtB5Pwq4.

References

Alexander, R. McNeill. 1982. *Locomotion of Animals*. London: Springer.

Bernstein, N. A. 2020. *Biomechanics for Instructors*. Translated by R. Whyman. Switzerland: Springer Nature.

Blazevich, A. 2017. *Sports Biomechanics: The Basics: Optimising Human Performance*. London: Bloomsbury.

Carrigan, C. R., and D. Gubbins. 1979. "The Source of the Earth's Magnetic Field." *Scientific American* 240, no. 2: 118–29.

Case, W. B., and M. A. Swanson. 1990. "The Pumping of a Swing from the Seated Position." *American Journal of Physics* 58: 463–67.

Chevedden, P. E., L. Eigenbrod, V. Foley, and W. Soedel. 1995. "The Trebuchet." *Scientific American* 273, no. 1: 66–71.

Collins, S. H., M. Wisse, and A. Ruina. 2001. "A Three-Dimensional Passive-Dynamic Walking Robot with Two Legs and Knees." *The International Journal of Robotics Research* 20: 607–15.

Collins, S. H., A. Ruina, R. Tedrake, and M. Wisse. 2005. "Efficient Bipedal Robots Based on Passive-Dynamic Walkers." *Science* 307: 1082–85.

Collins, S. H., P. G. Adamczyk, and A. D. Kuo. 2009. "Dynamic Arm Swinging in Human Walking." *Proceedings of the Royal Society B* 276: 3679–88.

Cooper, A., et al. 2022. "A Global Environmental Crisis 42,000 Years Ago." *Science* 371: 811–18.

Coppa, A., L. Bondioli, A. Cucina, D. W. Frayers, C. Jarrige, J-F. Jarrige, G. Quivron, M. Rossi, M. Vidale, and R. Macchiarelli. 2006. "Early Neolithic Tradition of Dentistry." *Nature* 440: 755–56.

Coriolis, G-G. 1835. "Sur les équations du mouvement relatif des systèmes de corps." *Journal de l'École Royale Polytechnique* 15: 144–54.

Crompton, R. H., T. C. Pataky, R. Savage, R. D'Aout, M. R. Bennett, M. H. Day, K. Bates, S. Morse, and W. I. Sellers. 2011. "Human-Like External Function of the Foot, and Fully Upright Gait, Confirmed in the 3.66 Million-Year-Old Laetoli Hominin Footprints by Topographic Statistics, Experimental Footprint-Formation and Computer Simulation." *Journal of the Royal Society Interface* 9: 707–19.

Davies, D. 2002. *A Brief History of Fighting Ships.* London: Robinson.

DeSilva, J. 2021. *First Steps: How Walking Upright Made Us Human.* London: William Collins.

Ennos, A.R. 2012a. *Solid Biomechanics.* Princeton: Princeton University Press.

Ennos, A. R. 2012b. "Balance, Angular Momentum and Sport." *Physics World* 25, no. 7: 21–26.

Ennos, A. R. 2020. *The Age of Wood: Our Most Useful Material and the Construction of Civilization.* New York: Scribner.

Farrell, J. W. 2020. *The Clock and the Camshaft.* Guilford, CT: Prometheus.

Fedor, J. V. 1961. *Theory and Design Curves for a Yo-Yo De-spin Mechanism for Satellites.* Technical Note D-708. Washington: National Aeronautics and Space Administration.

Ferrel, W. 1856. "An Essay on the Winds and Currents of the Ocean." *Nashville Journal of Medicine and Surgery* xi, nos. 4 and 5: 7–19.

Feynman, R. 1964. "Rotation in Space." *The Feynman Lectures on Physics*, vol. 1, ch. 20. https://feynmanlectures.caltech.edu/I_20.html.

Fleagle, J. 1974. "Dynamics of a Brachiating Siamang [*Hylobates (Symphalangus) syndactylus*]." *Nature* 248: 259–60.

Gbur, G. J. 2019. *Falling Felines and Fundamental Physics.* New Haven, CT: Yale University Press.

Goren-Inbar, N., M. Freikman, Y. Garfinkel, N. A. Goring-Morris, and L. Grosman. 2012. "The Earliest Matches." *PLos ONE* 7, no. 8: e42213.

Gould, D. W. 1975. *The Top.* Folkestone, UK: Bailey Brothers and Swinfen Limited.

Gribbin, J., and M. Gribbin. 2017. *Out of the Shadow of a Giant: How Newton Stood on the Shoulders of Hooke and Halley.* London: William Collins.

Harmand, S., et al. 2015. "3.3-Million-Year-Old Stone Tools from Lomekwi 3, West Turkana, Kenya." *Nature* 521: 310–15.

Hill, D. R. 1984. *A History of Engineering in Classical and Medieval Times.* New York: Barnes and Noble.

Hodges, H. 1970. *Technology in the Ancient World.* London: Allen Lane.

Hooke, R. 2007. *Micrographia, or Some Physiological Descriptions of Minute Bodies.* New York: Cosimo.

Jones, D. E. H. 1970. "The Stability of the Bicycle." *Physics Today* 23, no. 4: 34–40.

Kant, I. 2012. "Universal Natural History and Theory of the Heavens or Essay on the Constitution and the Mechanical Origin of the Whole Universe According to Newtonian Principles (1755)." In *Kant: Natural Science,* ed. E. Watkins. Cambridge, UK: Cambridge University Press, 182–308.

Keeley, L., and N. Toth. 1981. "Microwear Polishes on Early Stone Tools from Koobi Fora, Kenya." *Nature* 293: 464–65.

Knudson, D. 2007. *Fundamentals of Biomechanics: Second Edition.* New York: Springer.

Kvadze, E., O. Bar-Yosef, A. Belfer-Cohen, E. Boaretto, N. Jakeli, Z. Matskevich, and T. Meshveliani. 2009. "30,000-Year-Old Flax Fibers." *Science* 325: 1359.

Laithwaite, E. 1980. *An Engineer Through the Looking Glass.* London: BBC Books.

Laithwaite, E., and W. Dawson. 1999. *Propulsion System.* US Patent 5,860,317.

Leadbeater, E. 2008. *Spinning and Spinning Wheels.* Princes Risborough, UK: Shire.

Leighton, R., and R. Feynman. 1992. *Surely You're Joking, Mr. Feynman: Adventures of a Curious Character: Adventures of a Curious Character as Told to Ralph Leighton.* New York: Vintage.

Luncz, L. V., D. R. Braun, J. Marreiros, M. Bamford, C. Zeng, S. S. Pacome, P. Junghenn, Z. Buckley, X. Yao, and S. Cavalho. 2022. "Chimpanzee Wooden Tool Analysis Advances the Identification of Percussive Technology." *iScience* 25: 105315.

McGeer, T. 1990. "Passive Dynamic Walking." *The International Journal of Robotics Research* 9: 68–82.

Mickelborough, J., M. L. van der Linden, R. C. Tallis, and A. R. Ennos. 2003. "Muscle Activity During Gait Initiation in Normal Elderly People." *Gait and Posture* 19: 50–57.

Milks, A., D. Parker, and M. Pope. 2019. "External Ballistics of Pleistocene Hand-Thrown Spears: Experimental Performance Data and Implications for Human Evolution." *Scientific Reports* 25: 820.

Newton, I. 1999. *The Principia: Mathematical Principles of Natural Philosophy.* Translation by I. B. Cohen and A. Whiteman, preceded by *A Guide to Newton's Principia.* Berkeley: University of California Press.

Nye, M. J. 1999. "Temptations of Theory, Strategies of Evidence: P. M. S. Blackett and the Earth's Magnetism, 1947–52." *British Journal of the History of Science* 32: 69–92.

Persson, A. 2001a. "Back to Basics: Coriolis: Part 1—What Is the Coriolis Force?" *Weather* 55: 165–70.

Persson, A. 2001b. "Coriolis: Part 3—The Coriolis Force on the Physical Earth." *Weather* 55: 234–39.

Persson, A. 2001c. "The Obstructive Coriolis Force (Coriolis Part 4)." *Weather* 56: 204–8.

Persson, A. 2001d. "The Coriolis Force and the Geostrophic Wind (Coriolis Part 5)." *Weather* 56: 267–72.

Persson, A. 2003. "Proving that the Earth Rotates: The Coriolis Force and Newton's Falling Apple (Coriolis Part 9)." *Weather* 58: 267–72.

Persson, A. 2005. "The Coriolis Effect: Four Centuries of Conflict Between Common Sense and Mathematics, Part I: A History to 1885." *History of Meteorology* 2: 1–24.

Persson, A. 2006. "Hadley's Principle: Understanding and Misunderstanding the Trade Winds." *History of Meteorology* 3: 17–42.

Roach, N. T., M. Venkadesan, M. J. Rainbow, and D. E. Lieberman. 2013. "Elastic Energy Storage in the Shoulder and High-Speed Throwing" in *Homo. Nature* 498: 483–87.

Robertson, B. G. 1970. *Sopwith, the Man and His Aircraft.* Bedford: Sidney Press.

Smith, N. 1975. *Man and Water: A History of Hydro-Technology.* London: Scribner.

Smith, N. 1980. "The Origins of the Water Turbine." *Scientific American* 242: 138–48.

Smith, N. 1981. "Scientific Work." In *John Smeaton FRS,* ed. A. W. Skempton. London: Thomas Telford, 35–58.

Somerville, M. 2012. *Mechanism of the Heavens.* London: Forgotten Books.

Taylor, G. I. 1922. "The Motion of a Sphere in a Rotating Liquid." *Proceedings of the Royal Society of London A* 102: 180–89.

Thieme, H. 1997. "Lower Palaeolithic Hunting Spears from Germany." *Nature* 385: 807–10.

Tocheri, M. W., C. M. Orr, M. C. Jacofsky, and M. W. Marke. 2008. "The Evolutionary History of the Hand Since the Last Common Ancestor of *Pan* and *Homo." Journal of Anatomy* 212: 544–62.

Tomonaga, S-I. 1977. *The Story of Spin.* Translated by T. Oka. Chicago: Chicago University Press.

Uglow, J. 2002. *The Lunar Men.* London: Faber and Faber.

Voltaire. 1759. *Candide ou l'Optimisme.*

Walsh, K. J., A. Morbidelli, S. N. Raymond, D. P. O'Brien, A. M. Mandell. 2011. "A Low Mass for Mars from Jupiter's Early Gas-Driven Migration." *Nature* 475: 206–9.

Westcott, D. 1999. *Primitive Technology: A Book of Earth Skills.* Salt Lake City: Gibbs-Smith.

Winter, D. A. 1995. "Human Balance and Posture Control During Standing and Walking." *Gait and Posture* 3: 193–214.

Young, R. W. 2003. "Evolution of the Human Hand: The Role of Clubbing and Throwing." *Journal of Anatomy* 202: 165–74.

Illustration Credits

p. xiii: Hugh Llewelyn/Courtesy of Wikimedia Commons
p. 32: Andrew Z. Colvin/Courtesy of Wikimedia Commons
p. 34: Émile Alglave & J. Boulard. *The Electric Light: Its History, Production, and Applications*, translated by T. O'Conor Sloan (New York: D. Appleton & Co., 1884): p.224, fig.142.
p. 46: Kelvinsong/Courtesy of Wikimedia Commons
p. 53 (left): Alamy
p. 53 (right): Altes Museum, Berlin/Courtesy of Wikimedia Commons
p. 77 (left): Marcus Cyron/Courtesy of Wikimedia Commons
p. 77 (right): Toni Pecararo/Courtesy of Wikimedia Commons
p. 80: Courtesy of the Metropolitan Museum of Art. Gift of Mrs. Russell Sage, 1909
pp. 82, 99: Science Photo Library
p. 84: Archimedes screw, illustration from Soviet encyclopedia, 1926/Alamy
p. 85: Johnny Greig/Alamy
pp. 87, 114, 117, 127: Jahobr/Courtesy of Wikimedia Commons
p. 89 (top): Hadidehghanpour/Courtesy of Wikimedia Commons
p. 89 (bottom): Roland Ennos
p. 97: Courtesy of the Board of Trustees of the Science Museum
p. 100: Kasim Asim/Courtesy of the Board of Trustees of the Science Museum
p. 102: Clem Rutter/Courtesy of Wikimedia Commons
p. 104: Copperplate engraving by Wilson Lowry after an Illustration by J. Farey/Alamy
p. 108: US Naval Institute

p. 110: Tony Hisgett/Courtesy of Wikimedia Commons
p. 118: Bernard S. Janse/Courtesy of Wikimedia Commons
p. 122: Alfred John West/Courtesy of Wikimedia Commons
p. 129: Dionysos 1970/Courtesy of Wikimedia Commons
p. 140 (top): Archives of Pearson Scott Foresman/Courtesy of Wikimedia Commons
p. 140 (bottom): Silberwolf/Courtesy of Wikimedia Commons
p. 146: Emoscopes/Courtesy of Wikimedia Commons
p. 229: Nikolai Alexandrovich Bernstein's *Biomechanics for Instructors* (1925), translated by Rose Whymen

Index

Page numbers in *italics* refer to illustrations. Page numbers beginning with 241 refer to notes.